T0257770

Biodiversity Enrichment: Ecology and Agriculture

Biodiversity Enrichment: Ecology and Agriculture

Edited by **Neil Griffin**

New York

Published by Callisto Reference,
106 Park Avenue, Suite 200,
New York, NY 10016, USA
www.callistoreference.com

Biodiversity Enrichment; Ecology and Agriculture
Edited by Neil Griffin

International Standard Book Number: 978-1-63239-092-9 (Hardback)

Printed in the United States of America.

Contents

             Permissions

             List of Contributors

Preface

A research approach on biodiversity has been undertaken in this book. The biodiversity of plants, animals, fungi and microbes has been discussed in detail, along with biodiversity and its impact on mental health. The complexity of the association has been approached from various angles, taking into account the interventions at various stages. A scientific approach has been adopted in this book which exhibits the inter-connectivity of all the three levels, emphasizing the need for conservation and protection of the systems if human existence is to continually benefit from it. This book is a valuable reference for researchers and students interested in biodiversity.

Significant researches are present in this book. Intensive efforts have been employed by authors to make this book an outstanding discourse. This book contains the enlightening chapters which have been written on the basis of significant researches done by the experts.

Finally, I would also like to thank all the members involved in this book for being a team and meeting all the deadlines for the submission of their respective works. I would also like to thank my friends and family for being supportive in my efforts.

Editor

Ecological Section

Floral and Avifaunal Diversity of Thol Lake Wildlife (Bird) Sanctuary of Gujarat State, India

Jessica P. Karia

Additional information is available at the end of the chapter

1. Introduction

Wetlands are the ecotonal or transitional zones between terrestrial and aquatic ecosystems where the water table is usually at or near the surface of the land, which is covered by the shallow water (Mitsch & Gosselink, 1986). Due to these characteristics, wetlands provide opportunities for adaptations to different plant and animal species with high diversity of life-forms. Thus wetlands are among the most biologically diverse and productive ecosystems on earth. Wetlands can further be classified by one or more of the following attributes: (a) at least periodically, the land supports hydrophytes, (b) the substrate is predominantly undrained hydric soil, and (c) the substrate is saturated with water or covered by shallow water at some time during the growing season each year.

As per the convention on Wetlands of International importance (RAMSAR) (1971) – Article 1.1: wetlands are "Areas of marsh, fen, and peat land or water whether natural or artificial, permanent or temporary with water, that is static or flowing, fresh, brackish or salt including areas of marine water the depth of which does not exceed 6 meters." Also according to Article 2.1: "[Wetlands] may incorporate riparian and coastal zones adjacent to the wetlands, and islands or bodies of marine water deeper than six meters at low tide lying within the wetlands".

The values of the World's wetlands are increasingly receiving due attention as they contribute to a healthy environment in many ways. They help to retain water during dry periods, thus keeping the water-table high and relatively stable. During periods of flooding, they act to reduce flood levels and to trap suspended solids and nutrients directly flowing into the lakes. The removal of such wetland ecosystems because of urbanization or other factors typically causes lake water quality to worsen. In addition, wetlands are important feeding, breeding, and drinking area for wildlife and provide a stopping place and refuge for waterfowl. As with any natural habitat, wetlands are important in supporting species diversity and have a complex and important food web. The recent millennium assessment of ecosystems puts freshwater biodiversity as the most threatened of all types of biodiversity.

The interaction of man with wetlands during the last few decades has been of concern largely due to the rapid population growth, accompanied by intensified industrial, commercial and residential development. Thereby leading to pollution of wetlands by domestic, industrial sewage, and agricultural run-offs as fertilizers, insecticides and feedlot wastes. The fact that wetland values are overlooked has resulted in threat to the source of these benefits. Apart from the above the absence of reliable and updated information and data on extent of wetlands, their conservation values and socioeconomic importance has greatly hampered development of policy, legislation and administrative interventions by the state.

Fortunately in the recent years, the wetlands have received a good deal of attention. It really started with the conference held in Ramsar in Iran in 1971 where the first listing of wetlands of international importance was made and the contracting parties agreed to take necessary steps to safeguard these wetlands for posterity. India, as one of the original signatories, has made impressive efforts in initiating work for conservation and management of wetlands.

2. Indian scenario

India by its unique geographical position and with its annual rainfall of over 130 cm, and its varied terrain, and climate ranging from the cold arid of Ladakh to the warm arid of Rajasthan, with a cost line of over 7500 km, with its major river systems and lofty mountain ranges has, no wonder, a wealth of wetlands.

In addition to the various types of natural wetlands, a large number of man-made wetlands also contribute to the faunal and floral diversity. These man-made wetlands, which have resulted from the needs of irrigation, water supply, electricity, fisheries and flood control, are substantial in numbers. The various reservoirs, shallow ponds and numerous tanks support wetland biodiversity and add to the countries wetland wealth.

It is estimated that freshwater wetlands alone support 20 per cent of the known range of biodiversity in India (Deepa & Ramachandra, 1999). Wetlands in India occupy 58.2 million hectares, including area under wet paddy cultivation (Directory of Indian Wetlands).

Of about 35 Protected Areas (PAs) of India, which have been specifically notified for bird conservation, seven are in Gujarat (Grimmett et al. 1998). The State also falls within the Indus flyway a route that extents along the Indus valley from Pakistan to northwest India. This flyway is highly used by birds migrating from their breeding grounds in the Palearctic realm (Grimmett et al. 1998). The World Conservation Union (IUCN), International Wetland and Waterfowl Research Bureau (IWRB) and Birdlife International have rated this passage as the fourth major bird migration flyway in the world (Grimmett et al. 1998).

Gujarat is the State where the wetlands cover 27.1 lakhs hectares, a sizable area out of the total geographical area of the State. Of the total wetland area, inland wetlands cover 7.7%

and coastal wetland covers 92.3%. In coastal wetlands maximum area is under tidal flats/mud flats and the main contribution is from Great and Little Rann of Kachchh i.e. 1,930,581 ha. Analysis of the natural and man-made categories of wetlands indicate that, of the coastal wetlands, only 1.83% (mainly salt pans) is man-made, while in case of inland wetlands man-made wetlands account for 76.39% area.

Thol is one such man-made inland wetland situated in Mehsana district which is one of the top food grain producing districts in Gujarat (Anno. 1975). This marks the presence of well developed irrigation system consisting of wells and irrigation tanks. Thol water body is irrigation tank originally constructed in 1912 by the Gayekwadi State Rulers, built to prevent erosion and flooding and to store rainwater for irrigation purpose (Vaghela, 1993). Initially the area was declared as "Game Reserve" vides Government notification dated 29th May 1986 by Forest and Environment Department. Later on, due to its popularity amongst the bird fraternity, the area was notified as Bird Sanctuary through the notification GVN-53-88-WLP-1386-162-V.2 dated 18th November, 1988 under Section 18 of Wildlife (Protection) Act, 1972 (Anno. 2001).

Thol lake Wildlife Sanctuary which is now known as Thol Bird Sanctuary (TBS), as a part of conservation and management of Thol wetlands the biodiversity was studied to implement the Action Plan of Thol lake wildlife (Bird) sanctuary. This information will be comprehensive for preparing the management plan of the Sanctuary.

3. Study area

Thol Bird Sanctuary is situated in Mehsana district of Gujarat state, India between 23° 15' to 23° 30' N latitudes and 72° 30' to 72° 45' E longitude. It is a shallow water reservoir situated 25 km northwest of Ahmedabad and most popular birding place near Ahmedabad from Nal Sarovar Bird Sanctuary which is about 50 km away. Geographically, Thol Wildlife Sanctuary falls in the Kadi taluka of Mehsana district, North Gujarat region. Kadi taluka is head quarter of the district which is just 22 km away from the Sanctuary (Figure 1).

3.1. Salient features

Thol water body occupies a total area of 699 ha (6.99 sq.km.) and its periphery is 5.62 km long. Thol wetland catchment area is spread within six villages i.e. Thol, Jethlaj, Adhana, Vayana, Chandanpura, Jhaloda, which spreads 55.95 sq.km. It has well-developed canal based irrigation system. There are four head regulators at the water body to control the flow of water. The canals and their distributaries / sub-distributaries are about 19.97 km long.

The catchment area of the water body which covers 320 sq.km is located to its north and north-east so the spread is from Kadi taluka of Mehsana district and Kalol taluka of Gandhinagar district. These areas have seven small or big industrial areas they are, Karoli, Saij, Wamaj, Kalol, Chhatral, Indrad and Rajpur (Information from INDEXTb, Industrial Extension Bureau, Gandhinagar). Water finds its way through a number of canals draining into the feeder canal located on the north to northeastern sides of the water body. Water is received through Eastern canal, Saij-Hajipur canal, Irana-Indrad-Wamaj canal, Hajipur-Piyaj canal, Eastern feeder at Saghan drain, and Jaspur canal at Thol water body.

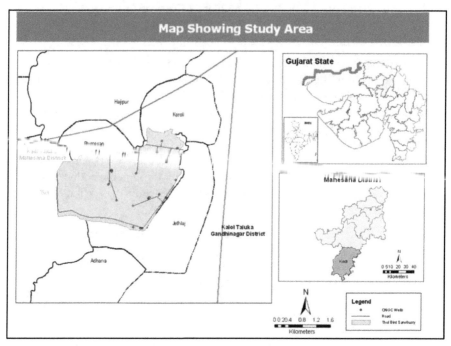

Figure 1.

In addition to the feeder canal, the water body receives run-off water directly from the catchment area. Before the feeder canal reaches the manmade wetland, there is a diversion, which is known as waste weir and is employed to control the volume of water in the water body. If the level of water reaches beyond 9 ft., the water is diverted to waste weir. Waste weir drains into a canal, which runs along the eastern boundary of the Thol pond/tank to reach Nalsarovar Bird Sanctuary located southwest of Thol Bird Sanctuary. Thol and Nalsarovar Bird Sanctuary are thus connected with each other.

There are no villages and settlements inside the sanctuary. Majority of the population is engaged in farming either as landholders or labourers. Also there are oil wells belonging to the public sector company Oil and Natural Gas Commission (ONGC) within the sanctuary area. There are total 21 number of wells among which 13 are functional. Polymer injection wells are 3 in number and Chase water wells are 5 in number. The total oil production from Thol area wells is 102 tpd.

3.2. Geology

Geologically, it is a part of the alluvial plain of recent age. The soil is clayey to sandy clay. There are no hard rock outcrops in and around the sanctuary.

3.3. Climate

Thol area experiences three distinct seasons namely winter (November to February), summer (April to May) and monsoon (June to September). Months of October and March mark the transition period from monsoon to winter and winter to summer respectively. The pond receives rainfall from July to September through the southwest monsoon. Old records for Mehsana district in general (Anno., 1975), as well as rainfall data of previous years at TS indicate that the rainfall is highly erratic and ranging from 189 to 786 mm.

4. Methodology

4.1. Land use / Cover studies

The methodology employed for preparation of Land use and land cover map included:

- Data collection
- Interpretation of satellite data
- Ground truth study
- Final map preparation

4.1.1. Data collection

- Downloading of Satellite imagery using the licensed software, Google Earth Pro having high resolution (<1.0m) data.
- Topographical maps as base map.
- Quick reconnaissance survey of the study area to get a feel of the entire ground area which can aid in the preliminary interpretation of the data.

4.1.2. Interpretation of satellite data

The downloaded satellite imagery was imported to Arc GIS 9.3 software and georeferencing of the imagery was done by registering it to the SOI maps through identification of common points between the map and the image. Considering the basic elements of interpretation such as tone, size, shape, texture, pattern, location, association, shadow, aspect and resolution, along with ground truth and ancillary information collected during the preliminary reconnaissance survey the interpretation was accomplished.

4.1.3. Ground truth study

A detailed ground truth was carried out to check the discrepancy of the interpreted data. It comprises of data collection of ground features along with the respective geographical position in terms of latitudes and longitudes.

4.1.4. Final map preparation

The interpreted file was then projected with Universal Transverse Mercator, which is universally followed projection system. The proportional presence of different land uses and land cover in terms of statistical percentages was derived for the study area. Appropriate legends were used to represent the various categories of land use and land cover, and were then written on the prepared land use and land cover maps. Based on interpreted map floral and faunal sampling site was selected so that the entire area will be covered.

4.2. Vegetation cover

The phyto-sociological studies were carried out using quadrant method with in terrestrial vegetation covered region. Quadrate plots were laid in triplicate at each selected locations. Density, frequency, abundance and dominance and their relative values were calculated along with IVI values (Ambasht, 1990). The basal area was calculated by formula using diameter at breast height (Ravlndianath & Premnath, 1997). Secondary analyses like different indices were calculated using this primary data (Odum, 1983).

The lower side of embankment had species diversity within this area the phytosociological studies were done. The grass cover region along the sanctuary boundary and on the beyts was surveyed and the herbs growing in this region was enlisted. The enlisting of the aquatic floral species like floating, emergent and submerged species had also been done.

4.3. Avifaunal studies

Avifaunal diversity studies sampling location was decided based on the water level and distribution as seen from interpreted satellite data. Observations were done by conducting field visits at regular intervals. Field works were conducted during winter season by visiting the place thrice in a season mainly from 0600 hr to 1200 hrs in the morning. The observations were recorded using field binocular (Pentax 10x50) and identified on basis of standard field guides like Grimmett et al. 1998, Salim Ali, 2002. This was done for both waterfowls and surrounding terrestrial birds. The bird diversity was classified according to its Order & Family, and their migratory statuses were noted.

4.4. Correlation between bird diversity & macrophytes

The relationship of the availability of bird diversity and macrophytes growing in the area was studied using statistical correlation method. The number of bird diversity distributed between the six sampling location and the available macrophytes diversity was documented.

5. Results

5.1. Land use / Cover studies

Visual interpretation of satellite data categorized area into five classes, they are shallow and deep water covered area, among terrestrial area it had been classified as vegetation cover, scrub land and agriculture land (Figure 2). The major portion of the sanctuary geographical area is covered by scrub land i.e. 36 per cent followed by 27 per cent of agricultural land. The category wise percentage area is as given in Table 1.

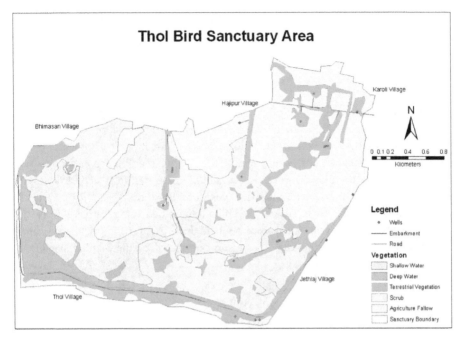

Figure 2.

Sr. No.	Class	Area (%)
1	Shallow water	15.08
2	Deep water	6.36
3	Vegetation Cover	15.69
4	Scrub	36.06
5	Agriculture	26.81

Table 1. Land use/cover Area Statistics of TBS

Each category is specific on its own as described below.

a. Shallow water: It could be delineated based on the light tone on the satellite data. Shallow water region was having less than 1 foot, as correlated on the ground. This was on the western side of the sanctuary boundary.

b. Deep water: It had dark coloured tone and smooth texture on the satellite data by which it was delineated as deep water. This was water filled region having more than 2 feet, as correlated on the ground. Its was on the western region or the corner of the sanctuary area.

c. Vegetation Cover: This was only 15 per cent of the area. This category will be described in details in following section.

d. Scrub: Most of the sanctuary area was covered by scrub, it is described as area coverage with less than 10 % of the canopy density (FSI, 2011) i.e. with scattered tree species and undergrowth dominated area.

e.   Agriculture: This was delineated based on the square patterns as seen on the data. The major crop grown in the region was Wheat, Juwar, and Bajra with water source as canal, bore or rainfed.

5.2. Vegetation cover

According to the vegetation map prepared using the satellite image of the Thol bird sanctuary area. There are three main patches of terrestrial vegetation first towards the Bhimasan village, on the north-east region along the water inlet to the Thol water body and lower side of the embankment. The vegetation towards the Bhimasan village was of monoculture type i.e. the plantation of *Acacia nilotica*, done by forest department. On the north east region along the water flow there was dominance of *Ipomoea fistula* and *Acacia nilotica* (Baval) vegetation. Lower side of the embankment had comparatively more species diversity where the phyto-sociological studies where done. Apart from this there were some patches of terrestrial vegetation which was again dominated by the planted species *Acacia nilotica* (Baval).

Phytosociological studies in the mixed vegetation type on the southern side of the sanctuary area showed presence of few species only. The highest abundance was of the *Acacia nilotica* (Baval) plantlet and its tree species had the highest IVI which showed that there was good regeneration of this species (Table 2). The understorey vegetation in this region was very less.

Sr. No.	Species	Abundance	IVI
1	*Acacia nilotica*	9.5	187.50
2	*Acacia planifrons*	2	22.80
3	*Zizyphus mauritiana*	2	19.67
4	*Azadirachta indica*	2	21.02

Table 2. Plant Species Status in Mixed Vegetation

The index calculated from the field data showed the dominance index to be greater than 0.5 indicating that one or two species contribute very highly in the community. Also from the high evenness index it could be judged that there is even distribution of the species (Table 3).

Index	Value
Simpson Dominance Index	0.67
Shannon-Wiener Diversity Index	1.03
Evenness Index	1.12
Species Richness Index	0.83

Table 3. Vegetation indices estimated from Mixed Vegetation

Along with this the enlisting of the species diversity within the sanctuary boundary was done, which showed the presence of 88 plant species (Annexure 1, see Appendix). It includes herbs, shrubs, grasses and hydrophytes species. There were in all 12 floating, emergent and submerged hydrophytes species.

5.3. Avifaunal diversity

Bird diversity recorded from Thol during present study were 144 in number out of it 76 are waterfowl rest are terrestrial birds (Annexure 2, see Appendix). The enlisted birds within the sanctuary area had 9 no. of rare and endangered species according to the Red Data Book (Table 4).

Within waterfowl there are members from 21 families and the family Anatidae had the most members i.e. 15, followed by Ardeidae having 11 members rest of the family have less then 10 members. This indicates that the ducks and geese are the dominating species, followed by herons, egrets and bitterns. Anatidae family members are mostly resident migratory or migratory species, only Comb duck is resident species. Resident migratory species are five i.e. Mallard, spot billed duck, bar headed goose, white eyed pochard and ruddy shelduck.

Terrestrial birds also had members from 21 numbers of families, and the Accipitridae family had highest number i.e. 10 members, followed by Cornidae having 9 members. These shows the dominating diversity within terrestrial area surrounding the water body are shikara, kite, eagle, vulture, buzzard, osprey and besra.

The statistics of residential status of species indicates that within aquatic birds diversity the highest number of species are resident-migratory i.e. 40 % while there are 33 % of resident species and 27 % of the migratory species. While within terrestrial birds species highest is of resident species having as high as 76 %, next is resident-migratory species of 21 % and just 3 % of migratory species.

Sr.No.	Scientific name	Common name	Migratory Status	Threatened Birds
1	*Anhinga melanogaster*	Oriental Darter	RM	NT
2	Mycteria leucocephala	Painted Stork	R	NT
3	*Phoenicopterus minor*	Lesser Flamingo	RM	NT
4	*Aythya nyroca*	White-eyed Pochard	RM	NT
5	*Threskiornis melanocephalus*	Oriental White Ibis	R	NT
6	*Grus antigone*	Sarus Crane	R	V
7	*Aquila heliaca*	Imperial Eagle	RM	V
8	*Aquila clanga*	Greater Spotted Eagle	RM	V
9	*Pelecanus philippensis*	Spotbilled Pelican	RM	V
NT - Near Threatened, V – Vulnerable, R – Residant, M – Migratory, RM- Resident-Migratory.				

Table 4. List of Threatened Birds in Thol Bird Sanctuary

The habitat requirement of the waterfowl inhabiting in Thol were studied. The details are as given in the Annexure 3. The foremost requirement is that of the open water both deep and shallow waters, in terms of percentage 47 % of birds which includes all the members of the dominating family Anatidae this habitat was used. Birds inhabiting in the muddy habitats are 22 % this includes some heron & egret, plovers, godwit, greenshank and sandpiper. Thereafter 16 % of the birds require emergent vegetation habitat, followed by 7 % shoreland, 6% agriculture and fallow land and 3 % of wooded area. Among the agriculture & fallow land habitat the dominating species is Sarus crane which comes under the vulnerable status. Among this there are overlapping of the use of habitat as per the birds resting, roosting and foraging habits.

5.4. Correlation between bird diversity & macrophytes

It has been observed that there was lot of variation in the floral and faunal diversity within selected six locations (Table 5). The sampling locations were selected based on the difference in the availability of water and the congregation of birds found in the region and the accessibility of the region. The sampling location P6 had presence of more floral diversity and location P1 had more of bird diversity.

Sample No.	Location	Longitude	Latitude	Remarks	Macrophyte Diversity	Bird Diversity
P1	Towards Bhimasan village	72°23′44.6″E	23°08′34.7″N	Shallow water nearly half foot	4	27
P2	Check Post side	72°23′35.0″E	23°08′25.8″N	Deep water around 2 feet	2	17
P3	Middle region	72°23′55.2″E	23°08′26.0″N	Shallow water	2	15
P4	Camp site	72°24′09.0″E	23°08′09.7″N	Muddy Area, no disturbance	2	16
P5	Towards Jetlaj village	72°24′50.3″E	23°07′52.6″N	Emergent Vegetation in pockets	4	11
P6	Towards ONGC well No. 30	72°24′41.0″E	23°08′13.6″N	Small ponds of water, gets flooded during monsoons, less biotic disturbance	6	23

Table 5. Floral and Faunal Diversity

6. Discussion

6.1. Land use / Cover studies

Land use and land cover classification is an essential prerequisite for any management operation as it is a direct indicator of the ecology of the area. This is particularly important to identify what kind of habitats in relation to the water level is formed and to find out habitat preferences of various species of waterfowl. Habitat is the natural home of any living form, may be an animal or a plant. Mc Farland (1980) suggested that birds respond to a summation of many factors and habitat selection thus, has some variability within a species. According to him a. characteristics of the terrain, b. nesting, feeding and drinking sites, c. food availability, d. other animals, are important factors influencing habitat selection. Therefore, identification of various habitat types are important factors influencing habitat selection.

The statistics reveals that there is more availability of shallow water habitat which in due course of time will be muddy region once water gets dried out. This study is very dynamic, thereby changes the types of birds visiting the wetlands. Also the availability water depends on the rainfall and the irrigation system. So this study needs to be conducted along with the bird census on regular basis.

6.2. Vegetation analyses

Macrophytes occurring in the area clearly indicate habitats and condition prevailing in the area of its occurrence. The habitat in the study area is mostly muddy and also it is saline, as indicated by sediment analysis (GEC, 2009).

Submerged rooted aquatic vegetation in Thol water body was of *Vallisneria spiralis* found near the Bhimasan village site and *Hydrilla verticillata* on the south eastern side near the ONGC well no. 30. The *Najas graminea* was found to be grown in abundant with *Hydrilla verticillata* and the *Potamogeton sp.* in the waters of Narmada Canal reaching TBS. Also on the sedges of the Narmada canal water there was growth of emergent hydrophyte *Typha* sp., and floating *Paspalum sp.* This indicates the presence of nutrients content in the Narmada canal water. While in sanctuary area waters such abundant growth of submerged hydrophytes was not seen. The bed of Thol water body was covered with the grass *Cynodon dactylon* (Darba) and the free-floating hydrophyte *Ipomea aquatic* (Nala ni Vel) on the south-western corner of the sanctuary. *Cynodon dactylon* (Darba) shows salt tolerance capacity and at the same time these are nutritious and palatable species. Rooted floating weeds *Nelumbo lutea* (Kamal) was seen to cover the small portion of the water body. On the check post side of the pond the growth of grass on the sedge was seen that of *Eragrostis sp.* Thus the reeds and sedges provide resting, rooting or nesting habitat for many species apart from providing an excellent cover, too many birds which take shelter in such habitats. In the middle portion of the sanctuary area from northern side, which is less disturbed site had the presence of free floating *Lemna* (Kaye) sp. on the edge the Amarantheceae member herb *Alternanthera sessilis*.

On the small bets, there appear mostly abundant *Ipomoea fistulosa* and in others bets *Acacia nilotica* (Desi baval) tree. On the other side of the waterbody i.e. on south eastern side near the ONGC well no. 30, there along with *Acacia nilotica, Parkinsonia accuminata* was found. These trees are extensively used by egrets, black ibis, crows, doves etc. for roosting. They are also used for nesting by crows, doves etc.

It was observed that on the south eastern side near the ONGC well no. 30 there appears reed meadow sedge, a seral stage where due to siltation, sedges, grasses grow abundantly. These include *Cyperus* sp. which play an important role of air circulation in the lake, as they are hollow and possess aerenchymous tissues. They help in gaseous exchanges of carbon dioxide and oxygen, which are thus made available to the submerged species. Also there was growth of *Polygonum sp.* on the sedges in this area. Emergent *Scirpus* sp. was also found in this region on the sedges, it has advantage to inhibit soil erosion and provide habitat for other wildlife. The plant rhizomes have medicinal value. So in this area we find diversity of macrophytes indicating the quality of water in Thol wetlands.

Major part of the Thol wetland sanctuary area is covered with scrub area. In this area there was sparse distribution of *Acacia nilotica* tree and the mesquite *Prosopis juliflora* (Gando baval) most of which are of shrubby appearance, seldom attaining a height of more that 3 meters. The ground is covered with grass *Cynodon dactylon* (Darba) and few herb species. The *Xanthium strumarium* (Gadariyu) an abnoxius weed appears at places along the shore and on some bets, which is indicative of excessive grazing in the area. This can be confirmed by the field survey in the area. The scrub area had more growth of herbs like the *Grangea maderaspatan* (zinki mundi), *Coldenia procumbens* (basario okharad), and *Glinus lotoides* (mitho okharad).

Among the tree species growing on the boundary of the Sanctuary were *Azadirachta indica* (Limdo), *Cordia myxa* (Gunda) and *Ailanthus excelsa* (Maharukh), etc. Apart from this on the southern boundary of the sanctuary area the natural vegetation grows where phytosociological parameters were studied. Micro level vegetational studies carried out aided to bring out sharp differences in the vegetation of these areas.

Each of the species within the community has a large measure of its structural and functional individualism and has more or less different ecological amplitude and modality (Singh and Joshi, 1979). This requires the understanding of the phytosociological status of each species within a community. Importance Value Index is a measure of plant status which brings out the overall role of a plant in a community (Ambasth, 1990). The study of phyto-sociology along with floristic composition proves useful in comparison of species from season to season and year to year (Singh, 1976). The study of vegetation its spatial distribution and analyses, and on field study indicates that the anthropogenic pressure had resulted in decrease in the undergrowth of the area. This would increases the possibility of the environmental stress i.e. soil erosion. This area shows the dominance of *Acacia nilotica* (Desi baval) with highest IVI of 187.5 and the dominance index. With the changing environmental conditions, the vegetation may reflect changes in structure, density and composition as observed by Gaur, (1982). The high evenness index shows the even distribution of vegetation in the community. It could be found out from survey that there is decrease in the undergrowth since it gets subjected to more anthropogenic pressure.

6.3. Avifaunal diversity

Bird communities are often referred as an ideal indicator to monitor the ecological condition of any wetlands as they impact on all the trophic levels of an aquatic ecosystem. On the other hand aquatic ecosystems have significant impact on migratory birds. Birds carry out, diverse ranges of ecological functions among vertebrates. As consumers, they help regulate populations of smaller animals they prey upon, disperse plant seeds, and pollinate flowering plants. As prey items, birds and bird eggs are consumed by a variety of larger predators.

Birds also benefit humans by providing important ecosystem services such as regulating services by scavenging carcasses and waste, by controlling population of invertebrates and vertebrate pests, by pollinating and dispersing the seeds of plants; and supporting services by cycling nutrients (Croll *et.al.*, 2005) and by contributing to soil formation (Post, 1998).

There are two birds which has been identified as flagship specis for Thol wetlands, being fresh water ecosystem, they are Sarus Crane (*Grus antigone*) and Osprey (*Pandion haliaetus*) since they represent the present ecosystem which is in need of conservation. They are distinctive in order to engender support and acknowledgement from the public.

- Sarus Crane (*Grus antigone*)

Sarus Crane is a large crane that is a resident breeding bird with *disjunct* populations that are found in parts of the Indian Subcontinent, Southeast Asia and Australia. Having height up to 1.8m, it is tallest of the flying birds; they are conspicuous and iconic species of open marshlands. As a species, the Sarus crane is classified as vulnerable this means that the global population has declined by about a third since 1980, and is expected to continue to do so until the late 2010. Estimates of the global population suggest that the population in 2000 was at best about 10% and at the worst just 2.5% of the numbers that existed in 1850 (BirdLife International, 2001). Unlike many cranes which make long migrations, the Sarus Crane does not; they may however make short-distance dispersal movements in response to rain or dry weather conditions. They tend to be more gregarious in the non-breeding season.

- Osprey (*Pandion haliaetus*)

Ospreys are sometimes known as the sea hawk, it is a large raptor, reaching 60 centimeters (24 in) in length with a 1.8 meter (6 ft) wingspan, is a resident-migratory species. They are widespread during winters in Indian Union, Bangladesh; Pakistan; Sri Lanka; Myanmar. Ospreys are diurnal, fish eating hawk, they flies up and down over the water scanning the surface for any fish coming up within striking depth.

Thol waterbody and surrounding area is most suitable habitat for Sarus, it can be appreciated from records that large number of Sarus congregations were seen. It has presence of over 50 birds feeding in the farmlands neighboring Thol, as late as 1998; the Sarus has remained the integral part of the avifauna of this territory (Singh & Tatu, 2000).

This shows that type of habitat is very important for wetland dependent species. Different species have different set of adaptations due to which they require certain types of habitats only. In case there is habitat loss in breeding areas it may directly result in loss of birds. Also the habitat is species specific and birds differ according to the habitat availability. Thus, the foremost requirement is identification of habitats in relation to various species of waterfowl.

TBS have variety of habitat which attracts many birds to the area. It was observed that dominating family Anatidae is having members like ducks and geese using open water habitat both deep and shallow. Thus the high usage of open water habitat explains why number of birds decrease with changes in the water spread and its level. Vijayan (1991) also reported preference of open water habitat over other categories by waterfowl at Keoladeo National Park. Anatidae group could truly be regarded as an indicator of the quality of habitat. As they depend on TBS for foraging, resting as well as roasting. Almost 60 per cent of Anatid members present are migratory and some species like the Whistling duck and Spot billed duck are potential breeders at TBS (GEER, 2002). Wetland could also be acting as a staging and dispersal area for the migrant ducks, which first arrive there and later spread to other smaller water bodies.

The migratory birds which come to TBS are coming mostly from northern and central Asia, Siberia and Europe or locally from Himalayas so their path is mostly north, north-east or north-west direction of TBS.

A total of 144 birds' species including 76 waterfowl and 68 terrestrial birds had been recorded at TBS during the study. The species diversity of waterfowl is similar to as recorded by Patel and Dharaiya, 2008 as 77 species. Species diversity was compared with other wetlands falling in semi-arid region like Wild Ass Sanctuary (Little Rann of Kachchh) and Nal Sarovar Bird Sanctuary. At Wild Ass Sanctuary (a seasonal fresh cum saline water protected wetland) Singh et al. (1999) had recorded 100 species of waterfowl (including wagtails and oriental pratincole) belonging to 18 families (as per old nomenclature). At Nal Sarovar Bird Sanctuary, Singh (1998) recorded about 113 waterfowl species. While Patel and Dharaiya in 2008 recorded 50 species of waterfowl. If we consider the area coverage Wild Ass Sanctuary is spread within 4953 sq. km., Nal Sarovar Bird Sanctuary spread within 120.82 sq. km. and TBS within 6.99 sq. km., if area is considered species diversity of TBS can be regarded remarkable.

The earlier study reveals that Nal Sarovar Bird Sanctuary which is just 50 km away, have high vegetation and faunal diversity compared to TBS (Patel, et al. 2006) due to different physical and hydrological configuration of largest natural fresh water reservoir. TBS and Nal Sarovar Bird Sanctuary are the valuable wetlands for migratory bird species. Moreover it can be also said that the Thol lake is more favored by the wetland obligatory birds. Since the study reveals that comparatively Nal Sarovar sanctuary had high disturbance score which indicates less healthy wetland for bird integrity than that of the TBS (Patel and Dharaiya, 2008). It was observed that towards the southwest direction of TBS is Nal Sarovar Bird Sanctuary, which is known to be one of the richest food crop (mainly paddy) growing areas, so there is continuous movement of birds between TBS and agricultural areas. This makes the birds visiting nearby village tanks and water bodies, which needs to be surveyed.

Thol has privilege of sustaining nine near threatened and vulnerable species. As reported by Chase et.al. (2000), presence of individual species may serve as indicator of the overall species composition of birds, but it may say less about the species richness, so the focus should be given to a diverse suite of the range of species representative for conservation purpose. The efforts should go in the line to conserve the threatened and lower risk species so that the population should not come down and they become extinct in near future. As per the red data guidelines they should be conserved when their populations are still healthy, before they become genetically impoverished and their populations gets fragmented. Out of nine vulnerable and near threatened species six are resident migratory species, and rests are resident species. Two species like eastern imperial eagle and greater spotted eagle are terrestrial birds and they are birds of prey.

6.4. Correlation between bird diversity & macrophytes

The enlisting of the bird diversity and availability of macrophytes in the region was subjected to statistical correlation which shows that there is positive correlation with 86 per cent of variance is related.

6.5. Avifaunal population trend in thol bird sanctuary

Avifaunal density trend was studied from the year 2000 to 2008. The year 2000 data was from the GEER foundation report 2002 while 2004 data was of Forest department; this was the first census of Thol bird sanctuary. Remaining two census data were taken of the year 2006 and 2008 conducted by Forest department. The trend changes in the population density of the birds found in the Thol bird sanctuary is as given in the table 6.

Sr. No.	Group of Species	2000[#]	2004[*]	2006[*]	2008[*]
1	Grebes	0	2	40	3
2	Pelicans	120	4	321	750
3	Cormorants & Darters	0	830	942	482
4	Herons & Egrets	1	479	485	210
5	Storks	4	83	236	95
6	Ibises & Spoonbills	19	768	183	5099
7	Flamingos	2	0	273	205
8	Geese & Ducks	419	1753	5599	7671
9	Cranes	525	380	664	1651
10	Reas, Crakes, Gallinules & Coots	0	21	943	552
11	Jacanas	0	0	0	0
12	Shorebirds & Waders	188	13839	8140	8120
13	Gulls, Terns & Skimmers	1	199	143	234
14	Kingfishers	0	10	15	25
15	Wagtails & Pipits	2	0	0	53
16	Eagles & Harriers	0	4	7	15
17	Total	1281	18372	17991	25165

GEER, (2002), * Forest Department Census

Table 6. Comparative Account of Birds Population (2000 to 2008)

It has been observed that over the years, there is increase in the population of the Pelicans, with sudden decrease in the year 2004. This growth can be attributed to the availability of the food; Pelicans mainly depend on the fish for food. It can be concluded that the forest department initiative of releasing fresh water fishes to the wetland was fruitful. Thus it could attract the migratory species to Thol wetlands.

The group Ibises & Spoonbill also shows the increase from just 19 numbers in year 2000 to 5,099 in the year 2008. The increase in population change of nil in year 2000 to 4,876 in year 2008 of Glossy ibis i.e. shore birds. The Gloosy Ibis requires the muddy habitat and they depend mainly on benthos for food. Thus over the years there is improvement of the food and availability of muddy habitat had increased Glossy ibis population. While, there is decrease in population of Eurasian spoonbill from 661 in 2004 to 187 in 2008. So it could be inferred that as there is habitat changes in the wetland ecosystem bird population changes. Reason could be that there is shift from water availability to muddy habitat availability.

Geese and Ducks group which had maximum diversity in Thol wetlands also shows the increasing trend from 2000 to 2008. This could is due to the increase in population of the migratory species Common teal to 4,769 in 2008. They are dependent on benthos as well as vegetation matter for food and require shallow water habitat.

Whereas the group Shorebirds & Waders shows the decreasing trend from highest of 13,839 in 2004 to 8,120 in 2008. This is largely because of decrease in migratory species Ruff from 13,345 (2004) to 5455 (2008), which is being compensated by the increase in population of Black tailed Godwit from 4 (2000) to 2,156 (2008). Ruffs are sporting birds they take larger quantities of weed seeds (Ali, 2002). Due to regulated supply of water for irrigation and developmental activities there is decrease in the agricultural fields and the availability of food for the species so there is negative change in the Ruff population. This year, Ruff species are not even noted, since as per the regulations due to construction work going on, the water supply was restricted causing the negligible population availability. This information was obtained from the forest officials of Thol wetlands.

Thus from the above discussion it can be concluded that due to adopted management practices there was overall increase in bird population of 1,281 (2000) to 25,165 (2008). But, definitely there was an overall change in the habitat causing the birds population to change accordingly. If we correlate the population of birds with the rainfall of the region it also had the increasing trend from 232 mm in 2000 to 786 mm in 2008 (Table 7). Rainfall in the year 2006 was slightly more as compared with 2008, but there was decrease in total bird population. This is probably due to favorable conditions prevailing in other wetlands also of the State during that period.

Looking at the avifaunal diversity it can be concluded that the Thol is the valuable wetlands for migratory bird species and it is more favored by the wetland obligatory birds because at Thol there is less human disturbance.

Sr. No.	Year	Rainfall (mm)
1	2008 - 2009	473
2	2007 - 2008	786
3	2006 - 2007	659
4	2005 - 2006	855
5	2004 - 2005	582
6	2003 - 2004	662
7	2002 - 2003	203
8	2001 - 2002	500
9	2000 - 2001	189
10	1999 - 2000	232

Table 7. Decadal Change In Rainfall Data of Thol.

7. Conclusion

TBS is important wetland of the western region as variety of migratory birds visit this wetland during winters. The study had identified the potential of TBS as an internationally important wetland due to species richness and home for nine near threatened and vulnerable species including endangered Sarus Crane, having pre-breeding congregations and nesting grounds.

It has been observed that though TBS is facing less human disturbance in comparison to Nal Sarovar Bird Sanctuary, there are certain threats if not controlled may increase. The foremost being the location of ONGC oil well within the sanctuary boundary and catchment area. It should be monitored regularly to check for oil spills or leaks as oil spills could be a threat for birds. Also the major portion of the sanctuary area is covered by agricultural region which is given to local people for cultivation at a meager rate. This activity causes disturbance to the birds. The withdrawal of water for irrigation which is through supply canals in command area and lift irrigation causes pressure to the wetland ecosystem.

Another major pressure on the Thol Bird Sanctuary is due to livestock population. Livestock of five peripheral villages as well as those belonging to the pastoral people from Kachchh and Saurashtra visit this area for grazing in scrub lands and for drinking water. The grazing pressure was confirmed by the field visit and the type of species growing in the region. The livestock includes goats, sheep, cows, buffaloes and camel which causes disturbance to birds. The forest department should manage TBS taking into consideration the mentioned threats.

Thus the present study has shown the importance of carrying out such a study on regular basis so as to monitor the changes of dynamic ecosystem due to concomitant changes in water regime at TBS. The study had a limited scope owing to its short span and was conceived only to document bird diversity. It is being suggested to carry out movement and dispersal pattern of migratory waterfowl. This can be extended to the neighboring villages' tank and water bodies which would enhance our knowledge about these winged visitors.

Author details

Jessica P. Karia

Enviro-GIS, Makarpura, Vadodara,Gujarat, India

Appendix

Sr. No.	Scientific Name	Vernacular Name
	Family: Mimosaceae	
1	*Acacia auriculiformis*	Pardeshi baval
2	*Acacia tomentosa*	Aniyar
3	*Acacia nilotica* ssp. *indica*	Baval
4	*Acacia farnesiana*	Talbaval
5	*Acacia planifrons*	
6	*Albizzia lebeck*	Siras
7	*Prosopis cineraria*	Khijado
8	*Prosopis glandulosa*	Gandobaval
9	*Prosopis juliflora*	Gandobaval
10	*Pithecellobium dulce*	Goras ambli
	Family: Malvaceae	
11	*Abutilon indica*	Kanski
	Family: Simaroubaceae	
12	*Ailanthus excelsa*	Maharukh
	Family: Amaranthaceae	
13	*Alternanthera sessilis*	
	Family: Meliaceae	
14	*Azadirachta indica*	Limdo
15	*Melia azadirach*	Bakan limdo
	Family: Ceasalpiniaceae	
16	*Bauhinia racemosa*	Asotri
17	*Cassia auriculata*	Aval
18	*Cassia fistula*	Garmalo
19	*Cassia italica*	Mindhi Aval
20	*Cassia occidentalis*	Kasundri
21	*Cassia siamea*	Kassod
22	*Cassia javanica renigera*	Pink cassia
23	*Cassia tora*	Kuvandio
24	*Delonix regia*	Gulmohar
25	*Tamarindus indica*	Ambli
	Family: Balanitaceae	
26	*Balanites aegyptiaca*	Ingorio
	Family: Papilionaceae	
27	*Butea monosperma*	Khakhro

Sr. No.	Scientific Name	Vernacular Name
28	*Crotalaria burhia*	Kharshan
29	*Derris indica*	Karanj
30	*Indigofera oblongifolia*	Zil, Ziladi
	Family: Poaceae	
31	*Cynodon dactylon*	Darba
32	*Cenchrus ciliaris*	Shukli
33	*Dicanthium annulatum*	Jinjavo
34	*Eragrostis sp.*	-
35	*Dactyolactelum aegypticum*	
	Family: Cyperaceae	
36	*Scirpus sp.*	
37	*Cyperus sp.*	-
	Family: Capparaceae	
38	*Capparis decidua*	Kerdo
39	*Capparis sepiara*	Kanther
	Family: Asclepiadaceae	
40	*Calotropis procera*	Nano akado
41	*Calotropis gigantea*	Akdo
	Family: Verbenaceae	
42	*Clerodendron multiflorum*	Arani
	Family: Boraginaceae	
43	*Coldenia procumbens*	Basario Okharad
44	*Heliotropium indicum*	Hathisundho
45	*Cordia myxa*	Gunda
	Family: Menispermaceae	
46	*Cocculus hirsutus*	Patalagarudi
	Family: Solanaceae	
47	*Datura metal*	Dholo Dhanturo
48	*Solanum xanthocarpum*	-
	Family: Euphorbiaceae	
49	*Euphorbia hirta*	Dudheli
50	*Euphorbia nivulia*	Thor
51	*Euphorbia obiculata*	-
52	*Phyllanthus reticulata*	Kamboi
53	*Ricinus communis*	Castor
	Family: Moraceae	
54	*Ficus benghalensis*	Vad
55	*Ficus religiosa*	Peepal
	Family: Asteraceae	
56	*Grangea maderaspatana*	Zinki Mundi
57	*Launaea procumbens*	Moti Bhonpatri
58	*Xanthium strumarium*	Gadariyu

Sr. No.	Scientific Name	Vernacular Name
	Family: Molluginaceae	
59	*Glinus lotoides*	Mitho Okharad
	Family: Ulmaceae	
60	*Holoptelea integrifolia*	Kanjo
	Family: Sterculiaceae	
61	*Helicteris isora*	Maradiya
	Family: Hydrocharitaceae	
62	*Hydrilla verticilliata*	-
63	*Vallisneria natans*	-
	Family:Convolvulaceae	
64	*Ipomoea fistulosa*	Naffatiyu
65	*Ipomoea aquatica*	Nada ni Vel
	Family:Lemnaceae	
66	*Lemna sp.*	Kaye
	Family: Sapotaceae	
67	*Madhuca indica*	Mahudo
	Family: Anacardiaceae	
68	*Mangifera indica*	Keri
	Family: Celastraceae	
69	*Maytenus emarginata*	Vicklo
	Family: Moringaceae	
70	*Moringa oliefera*	Saragavo
	Family: Najadaceae	
71	*Najas graminea*	-
	Family: Nelumbonaceae	
72	*Nelumbo lutea*	Kamal
	Family: Rubiaceae	
73	*Oldenlandia sp.*	-
	Family: Fabaceae	
74	*Parkinsonia floridum*	
	Family: Potamogetonaceae	
75	*Potamogeton sp.*	
	Family: Annonaceae	
76	*Polyalthia longifolia*	Asopalav
	Family: Polygonaceae	
77	*Polygonum plebeium*	-
78	*Polygonum glabrum*	-
	Family: Salvadoraceae	
79	*Salvadora persica*	Piludi
80	*Salvadora olioedis*	Pilu

Sr. No.	Scientific Name	Vernacular Name
	Family: Verbenaceae	
81	*Tectona grandis*	Sag
	Family: Zygophyllaceae	
82	*Tribulus terrestris*	Bethu Gokhru
	Family: Typhaceae	
83	*Typha angustata*	
	Family: Rhamnaceae	
84	*Zizyphus mauritiana*	Bordi
85	*Zizyphus nummularia*	Chani Bor

Annexure 1. List of Vegetation (Aquatic & Terrestrial) Recorded in Thol Bird Sanctuary

Sr. No.	Scientific name	Common name	Migratory Status	Food Habits
AQUATIC BIRDS				
	Order: Anseriformes			
	Family: Anatidae			
1	*Anas acuta*	Northern Pintail	M	Aquatic plants,grains, insects, tadpoles etc.
2	*Anas clypeata*	Northern Shoveler	M	Water insects, snails, planktons, fish spawn.
3	*Anas crecca*	Common Teal	M	Chiefly vegetable matter, insects, crustaceans etc
4	*Anas penelope*	Eurasian Wigeon	M	Largerly vegetarian
5	*Anas platyrhynchos*	Mallard	RM	Largerly vegetarian
6	*Anas poecilorhyncha*	Spot billed Duck	RM	Chiefly vegetable matter
7	*Anas querquedula*	Garganey	M	Largerly vegetarian
8	*Anas strepera*	Gadwall	M	Largerly vegetarian
9	*Anser anser*	Greylag Goose	M	Vegeterian, winter crops, grass, aquatic weeds
10	*Anser indicus*	Bar-headed Goose	RM	Chiefly green shoots of winter crops - wheat/gram
11	*Aythya ferina*	Common Pochard	M	Vegetable matter, insects, molluscs, small fish etc
12	*Aythya nyroca*	White-eyed Pochard	RM	Vegetable matter, insects, molluscs, small fish etc
13	*Sarkidiornis melanotos*	Comb Duck	R	Grain, shoots vegetable matter
14	*Tadorna ferruginea*	Ruddy Shelduck	RM	Vegetable matter, insects, molluscs, small fish etc
15	*Tadorna tadorna*	Common Shelduck	M	Ominivorous, molluscs, algae, seeds etc.

Sr. No.	Scientific name	Common name	Migratory Status	Food Habits
	Family: Dendrocygnidae			
16	*Dendrocygna javanica*	Lesser Whistling Duck	R	Largerly vegetarian - shoots and grain.
	Family: Anhingidae			
17	*Anhinga melanogaster*	Oriental Darter	RM	Fish
	Family: Ardeidae			
18	*Ardea cinerea*	Grey Heron	RM	
19	*Ardea purpurea*	Purple Heron	RM	Fish, Frogs, snakes etc.
20	*Ardeola grayii*	Indian Pond Heron	R	Frogs, fish, crabs and insects
21	*Bubulcus ibis*	Cattle Egret	R	Chiefly grasshoppers, blue bottle flies,lizards, fish etc
22	*Casmerodius albus*	Great Egret	RM	Fish, Frogs,etc.
23	*Egretta garzetta*	Little Egret	R	Insects, fish, frogs etc.
24	*Egretta gularis*	Western Reef Egret	RM	Mainly crustaceans, molluscs and fish
25	*Ixobrychus minutus*	Little Bittern	RM	Fish, molluscs etc.
26	*Ixobrychus sinensis*	Yellow Bittern	RM	Fish, frogs, molluscs etc.
27	*Mesophoyx intermedia*	Intermediate Egret	RM	Fish, frogs etc.
28	*Nycticorax nycticorax*	Black crowned Night Heron	R	Crabs, fish, frogs, aquatic insects, etc.
	Family: Charadriidae			
29	*Charadrius alexandrinus*	Kentish Plover	RM	Insects and crustacea
30	*Charadrius dubius*	Little Ringed Plover	RM	Insects, sand-hoppers, tiny crabs, etc.
31	*Vanellus indicus*	Red wattled Lapwing	R	Insects, grubs, molluscs, etc.
32	*Vanellus leucurus*	White tailed Lapwing	M	Aquatic insects and other vertebrates
33	*Vanellus malabaricus*	Yellow wattled Lapwing	R	Insects, grubs, molluscs, etc.
34	*Calidris minuta*	Little Stint	M	Tiny insects, crustaceans and molluscs.
	Family: Recurvirostridae			
35	*Himantopus himantopus*	Black winged Stilt	RM	Worms, molluscs, aquatic insects, etc.
36	*Recurvirostra avosetta*	Pied Avocet	RM	Worms, aquatic insects and small crustacea, etc.

Sr. No.	Scientific name	Common name	Migratory Status	Food Habits
	Family: Ciconiidae			
37	*Anastomus oscitans*	Asian Openbill	R	Frogs, crabs, large insects and other small living things.
38	*Ciconia episcopus*	Woolly necked Stork	R	Fish. Frogs. Reptiles, crabs, molluscs, large insects, etc.
39	*Mycteria leucocephala*	Painted Stork	R	Fish, frogs and snakes.
	Family: Jacanidae			
40	*Metopidius indicus*	Bronze Winged Jacana	R	Seeds, roots, etc., aquatic plants, insects and molluscs
41	*Hydrophasianus chirurgus*	Pheasant-tailed Jacana	R	Seeds, roots, etc., aquatic plants, insects and molluscs
	Family: Laridae			
42	*Chlidonias hybridus*	Whiskered Tern	RM	Tiny fishes, tadpoles, crabs, grasshoppers and insects.
43	*Sterna albifrons*	Little Tern	R	Small fish, crustaceans, insects.
44	*Sterna aurantia*	River Tern	R	Fish, crustaceans, tadpoles and water insects.
	Family: Pelecanidae			
45	*Pelecanus philippensis*	Spotbilled Pelican	RM	Fish
46	*Pelecanus crispus Bruch*	Great White Pelican	M	Fish, crustaceans
	Family: Phalacrocoracidae			
47	*Phalacrocorax carbo*	Great Cormorant	RM	Almost exclusively fish
48	*Phalacrocorax fuscicollis*	Indian Cormorant	RM	Almost exclusively fish
49	*Phalacrocorax niger*	Little Cormorant	RM	Exclusively fish
	Family: Phoenicopteridae			
50	*Phoenicopterus ruber*	Greater Flamingo	RM	Crustaceans, worms, insect larvae, seeds of marsh plants.
51	*Phoenicopterus minor*	Lesser Flamingo	RM	Phytoplankton (algae, diatoms, etc.)
	Family: Podicipedidae			
52	*Tachybaptus ruficollis*	Little Grebe	RM	Aquatic insects and larvae, tadpoles, etc.
	Family: Scolopacidae			
53	*Limosa limosa*	Black tailed Godwit	M	Worms, molluscs, crabs, insects.
54	*Limosa lapponica*	Bar-tailed Godwit	M	Marine invertebrates, insects.

Sr. No.	Scientific name	Common name	Migratory Status	Food Habits
55	*Tringa glareola*	Wood Sandpiper	M	Insects, larvae, worms and molluscs.
56	*Tringa hypoleucos*	Common Sandpiper	RM	Insects, worms , molluscs.
57	*Tringa nebularia*	Common Greenshank	M	Insects and other invertebratres, tadpoles, frogs.
58	*Tringa ochropus*	Green Sandpiper	M	
59	*Tringa stagnatilis*	Marsh Sandpiper	M	Insects, invertebrates and small frogs.
	Family. Threskiornithidae			
60	*Platalea leucorodia*	Eurasian Spoonbill	RM	Tadpoles, frogs, molluscs, insects and vegetable matter
61	*Plegadis falcinellus*	Glossy Ibis	RM	Molluscs, crustaceans, insects, etc.
62	*Pseudibis papillosa*	Red-naped/Black Ibis	R	Insects, grain and small reptiles.
63	*Threskiornis melanocephalus*	Oriental White Ibis	R	Tadpoles, frogs, molluscs, insects and vegetable matter
	Order: Coraciiformes			
	Family: Alcedinidae			
64	*Alcedo atthis*	Small Blue Kingfisher	R	Small fish, tadpoles and aquatic insects.
	Family: Cerylidae			
65	*Ceryle rudis*	Pied Kingfisher	R	Fish, tadpoles, frogs and aquatic insects.
	Family: Dacelonidae			
66	*Halcyon smyrnensis*	White-breasted Kingfisher	R	Fish, tadpoles, lizard, grasshoppers and other insects
	Order: Gruiformes			
	Family: Gruidae			
67	*Grus antigone*	Sarus Crane	R	Grain, shoots and other vegetable matter, insects, reptiles.
68	*Grus grus*	Common Crane	M	Largerly vegetarian, tubers, grain, insects and small reptiles
69	*Grus virgo*	Demoiselle Crane	M	
	Family: Rallidae			
70	*Amaurornis akool*	Brown Crake	R	
71	*Amaurornis phoenicurus*	White-breasted Waterhen	R	Insects, worms, molluscs, grain, etc.
72	*Fulica atra*	Common Coot	RM	Grass and Paddy shoots, aquatic weeds, insects, etc.

Sr. No.	Scientific name	Common name	Migratory Status	Food Habits
73	*Gallicrex cinerea*	Watercock	R	Largely vegetarin - seeds and green shoots of rice etc.
74	*Gallinula chloropus*	Common Moorhen	RM	Insects, worms, molluscs, grain, etc.
75	*Porphyrio porphyrio*	Purple Swamphen/Moorhen	R	Shoots and vegetable matter, insects and molluscs.
	Family: Accipitridae (Or.Ciconiformese)			
76	*Circus aeruginosus*	Western Marsh Harrier	M	Frogs, fish small birds, mammals and carrion.
	TERRESTRIAL BIRDS			
	Order: Apodiformes			
	Family: Apodidae			
77	*Apus nipalensis*	House Swift	RM	Chiefly dipterous insects.
78	*Cypsiurus balasiensis*	Asian Palm-Swift	R	Tiny winged insects.
	Order: Ciconiiformes			
	Family: Accipitridae			
79	*Accipiter badius*	Shikra	R	Lizards, mice, squirrels, birds etc.
80	*Accipiter virgatus*	Besra	R	Largely small birds, mice, bats, lizards and insects.
81	*Aquila heliaca*	Imperial Eagle	RM	Rodents, ground dwelling birds, reptiles, etc.
82	*Aquila clanga*	Greater Spotted Eagle	RM	Frogs, waterfowl, small birds, etc.
83	*Elanus caeruleus*	Black-shouldered Kite	R	Locusts, crickets, mice, lizards,etc.
84	*Milvus migrans*	Black Kite	R	Offal and garbage, earthworms, mice, lizards etc.
85	*Neophron percnopterus*	Egyptian Vulture	RM	Animal carcasses and freshwater turtles.
86	*Pandion haliaetus*	Osprey	RM	Fish
87	*Pernis ptilorhyncus*	Oriental Honey-buzzard	RM	Honeybee larvae, small birds, reptiles, frogs etc.
88	*Spilornis cheela*	Crested Serpent-Eagle	R	Frogs, lizards, rats, snakes,etc.
	Order: Columbiformes			
	Family: Columbidae			
89	*Columba livia*	Rock Pigeon	R	Cereals, pulses, groundnuts,etc.
90	*Streptopelia chinensis*	Spotted Dove	R	
91	*Streptopelia decaocto*	Eurasian	R	

Sr. No.	Scientific name	Common name	Migratory Status	Food Habits
		Collared-Dove		
92	*Streptopelia orientalis*	Oriental Turtle-Dove	RM	Paddy, cereals, bamboo and grass seeds.
93	*Streptopelia tranquebarica*	Red Collared-Dove	R	
94	*Treron phoenicoptera*	Yellow-footed Green-Pigeon	R	Fruits and berries.
	Order: Coraciiformes			
	Family: Coraciidae			
95	*Coracias benghalensis*	Indian Roller	R	Insects.
96	Merops orientalis	Little Green Bee-eater	R	Insects, chiefly diptera and hymenoptera
97	*Centropus sinensis*	Greater Coucal	R	caterpillars, large insects, snails, lizards young mice etc.
	Family: Cuculidae			
98	*Cuculus micropterus*	Indian Cuckoo	RM	Mainly caterpillars, insects, etc.
99	*Eudynamys scolopacea*	Asian Koel	R	Largely fruits and berries, caterpillars and insects.
	Family: Phasianidae			
100	*Coturnix coturnix*	Common Quail	RM	Grain and grass seeds, termites, etc.
101	*Francolinus pictus*	Painted Francolin	R	Grain, grass seeds, green shoots, white ants and insects.
102	*Francolinus pondicerianus*	Grey Francolin	R	Grain, seeds, termites , beetle larvae, etc.
103	*Pavo cristatus*	Indian Peafowl	R	Grain, Vegetable shoots, insects, lizards, snakes, etc.
	Order: Passeriformes			
	Family: Aegithalidae			
104	*Aegithalos leucogenys*	White-cheeked Tit	R	
	Family: Alaudidae			
105	*Eremopterix grisea*	Ashy-crowned Sparrow-Lark	R	Seeds and insects.
106	*Eremopterix nigriceps*	Black-crowned Sparrow-Lark	R	
	Family: Cisticolidae			
107	*Prinia inornata*	Plain Prinia	R	Insects, caterpillars, ants, small beetles,etc.
108	*Prinia socialis*	Ashy Prinia	R	Insects.
	Family: Corvidae			
109	*Aegithina tiphia*	Common Iora	R	Insects, their eggs and larvae.

Sr. No.	Scientific name	Common name	Migratory Status	Food Habits
110	*Corvus splendens*	House Crow	R	Offal, dead sewe rat, kitchen scraps and refuse, termitesetc
111	*Dendrocitta vagabunda*	Rufous Treepie	R	Fruits, insects, lizards, frogs, centipedes etc.
112	*Dicrurus leucophaeus*	Ashy Drongo	RM	Mainly insects, occasionally reptiles, and small birds.
113	*Dicrurus macrocercus*	Black Drongo	R	Insects, flower nectar, occasionally small birds.
114	*Garrulus glandarius*	Eurasian Jay	R	
115	*Pericrocotus cinnamomeus*	Small Minivet	R	Insects and their larvae.
116	*Rhipidura albicollis*	White-throated Fantail	R	
117	*Rhipidura aureola*	White-browed Fantail	R	Insects, chiefly diptera and hemiptera.
	Family: Hirundinidae			
118	*Delichon urbica*	Northern House-Martin	RM	Midges and other insects.
119	*Hirundo smithii*	Wire-tailed Swallow	R	Midges
	Family: Laniidae			
120	*Lanius vittatus*	Bay-backed Shrike	R	Locusts,lizards , large insects, etc.
	Family: Muscicapidae			
121	*Copsychus saularis*	Oriental Magpie-Robin	R	Insects, flower nectar of Salmalia and Erythrina.
122	*Saxicoloides fulicata*	Indian Robin	R	Insects and their eggs, spiders, etc.
	Family: Nectariniidae			
123	*Nectarinia asiatica*	Purple Sunbird	R	Insects and spiders, very largely flower nectar.
	Family: Passeridae			
124	*Anthus campestris*	Tawny Pipit	RM	
125	*Anthus rufulus*	Paddyfield Pipit	R	Weev and other small insects
126	*Lonchura striata*	White-rumped Munia	R	Grass seeds, etc.
127	*Motacilla cinerea*	Grey Wagtail	M	Tiny insects.
128	*Motacilla flava*	Yellow Wagtail	RM	Insects, spiders and invertebrates, etc.
129	*Passer domesticus*	House Sparrow	R	Grains, insects, fruit buds, flower nectar, etc.
	Family: Pycnonotidae			
130	*Pycnonotus cafer*	Red-vented	R	Insects, fruits and berries, peas

Sr. No.	Scientific name	Common name	Migratory Status	Food Habits
		Bulbul		and vegetables etc.
131	*Pycnonotus leucotis*	White-eared Bulbul	R	Kitchen scraps, berries of peelu and wild caper.
	Family: Sturnidae			
132	*Acridotheres ginginianus*	Bank Myna	R	Grasshoppers and other insects.
133	*Acridotheres tristis*	Common Myna	R	Fruits, insects, kitchen scraps, etc.
134	*Sturnus pagodarum*	Brahminy Starling	R	Chiefly berries, wild figs and insects.
135	*Sturnus roseus*	Rosy Starling	M	Locusts, berries, nectar of Salmalia, etc.
	Family: Sylviidae			
136	*Acrocephalus arundinaceus*	Great Reed Warbler		
137	*Orthotomus sutorius*	Common Tailorbird	R	Tiny insects, their eggs and grubs, flower nectar.
138	*Turdoides caudatus*	Common Babbler	R	Insects, berries, grain and flower nectar.
139	*Turdoides earlei*	Striated Babbler	R	Insects, snails and some vegetable matter.
140	*Turdoides malcolmi*	Large Grey Babbler	R	Insects, berries, grain and flower nectar.
141	*Turdoides striatus*	Jungle Babbler	R	Spiders, cockaroaches, insects and their larvae grain, etc.
	Order: Psittaciformes			
	Family: Psittacidae			
142	*Psittacula krameri*	Rose ringed parakeet	R	Ripening fruits, standing crops of maize and jowar.
	Order: Strigiformes			
	Family: Strigidae			
143	*Athene brama*	Spotted Owlet	R	Chiefly beetle and other insects, mice, lizards, etc.
	Order: Upupiformes			
	Family: Upupidae			
144	*Upupa epops*	Eurasian Hoopoe	RM	Insects, grubs and pupae.

Annexure 2. List of Aquatic & Terrestrial Birds in Thol Bird Sanctuary

Sr. No.	Scientific name	Common name	Migratory Status	Open Water - Deep	Open Water - Shallow	Emergent Aquatic Vegetation	Muddy	Shore Land	Surrounding Environment - Agri	Surrounding Environment - Fallow	Wooded Areas
	Order: Anseriformes										
	Family: Anatidae										
1.	Anas acuta	Northern Pintail	M	L	H	L	M				
2.	Anas clypeata	Northern Shoveler	M		H	M	M				
3.	Anas crecca	Common Teal	M		H	M					
4.	Anas penelope	Eurasian Wigeon	M	L	H	L	L				
5.	Anas platyrhynchos	Mallard	RM	L	H	L	L				
6.	Anas poecilorhyncha	Spot billed Duck	RM	L	H	L	L				
7.	Anas querquedula	Garganey	M	L	H						
8.	Anas strepera	Gadwall	M		M	L	L	M	H	H	
9.	Anser anser	Greylag Goose	M		M	L	L	M	H	H	
10.	Anser indicus	Bar-headed Goose	RM	H	M	L	L				
11.	Aythya ferina	Common Pochard	M	H	H						
12.	Aythya nyroca	White-eyed Pochard	RM	L	M	M	L				
13.	Sarkidiornis melanotos	Comb Duck	R	L	H		M				
14.	Tadorna ferruginea	Ruddy Shelduck	RM	L	M	L	L				
15.	Tadorna tadorna	Common Shelduck	M	L	L	L	L	L			
	Family: Dendrocygnidae										
16.	Dendrocygna javanica	Lesser Whistling Duck/Teal	R	L	H	M	L	L	L	L	L
	Family: Anhingidae										
17.	Anhinga melanogaster	Oriental Darter	RM	H	M	M		L			
	Family: Ardeidae										
18.	Ardea cinerea	Grey Heron	RM			L	H	M			
19.	Ardea purpurea	Purple Heron	RM			H	M	L			
20.	Ardeola grayii	Indian Pond Heron	R			L	H	M	H	H	
21.	Bubulcus ibis	Cattle Egret	R			M	M	H	L	L	
22.	Casmerodius albus	Great Egret	RM		H	M	M	L	M	L	
23.	Egretta garzetta	Little Egret	R		M	M	H	M	M	M	M
24.	Egretta gularis	Western Reef Egret	RM		H	M	H	H	M	M	M
25.	Ixobrychus minutus	Little Bittern	RM			H	L	H	L	L	M
26.	Ixobrychus sinensis	Yellow Bittern	RM			H	H		L		
27.	Mesophoyx intermedia	Intermediate Egret	RM			H	H		L		
28.	Nycticorax nycticorax	Black crowned Night Hero	R			M	M	M	L		
	Family: Charadriidae										
29.	Charadrius alexandrinus	Kentish Plover	RM			L	H	M			
30.	Charadrius dubius	Little Ringed Plover	RM				H	M	M	M	
31.	Vanellus indicus	Red wattled Lapwing	R				L	H	M	M	M
32.	Vanellus leucurus	White tailed Lapwing	M				L	H	M	M	M
33.	Vanellus malabaricus	Yellow wattled Lapwing	R				L	H	M	M	M
34.	Calidris minuta	Little Stint	M			L	H	L	L	L	
	Family: Recurvirostridae										
35.	Himantopus himantopus	Black winged Stilt	RM		M	M	H	M	M	M	
36.	Recurvirostra avosetta	Pied Avocet	RM		M	L	H		L	L	
	Family: Ciconiidae										
37.	Anastomus oscitans	Asian Openbill	R		H	L	L		M	M	M
38.	Ciconia episcopus	Woolly/White necked Stork	R		H	L	M		L	L	L
39.	Mycteria leucocephala	Painted Stork	R		H		M				
	Family: Jacanidae										
40.	Metopidius indicus	Bronze winged Jacana	R		M	H	L	L			
41.	Hydrophasianus chirurgus	Pheasant-tailed Jacana	R		M	H	L				
	Family: Laridae										
42.	Chlidonias hybridus	Whiskered Tern	RM	H	H		L	M			M
43.	Sterna albifrons	Little Tern	R	H	H		L	M			L

No.	Scientific name	Common name	Status
44	*Sterna aurantia*	River Tern	R
	Family: Pelecanidae		
45	*Pelecanus crispus* Bruch	Great White Pelican	RM
46	*Pelecanus philippensis*	Spot-billed Pelican	M
	Family: Phalacrocoracidae		
47	*Phalacrocorax carbo*	Great Cormorant	RM
48	*Phalacrocorax fuscicollis*	Indian Cormorant	RM
49	*Phalacrocorax niger*	Little Cormorant	RM
	Family: Phoenicopteridae		
50	*Phoenicopterus ruber*	Greater Flamingo	RM
51	*Phoenicopterus minor*	Lesser Flamingo	RM
	Family: Podicipedidae		
52	*Tachybaptus ruficollis*	Little Grebe	RM
	Family: Scolopacidae		
53	*Limosa limosa*	Black-tailed Godwit	M
54	*Limosa lapponica*	Bar-tailed Godwit	M
55	*Tringa glareola*	Wood Sandpiper	M
56	*Tringa hypoleucos*	Common Sandpiper	RM
57	*Tringa nebularia*	Common Greenshank	M
58	*Tringa ochropus*	Green Sandpiper	M
59	*Tringa stagnatilis*	Marsh Sandpiper	M
	Family: Threskiornithidae		
60	*Platalea leucorodia*	Eurasian Spoonbill	RM
61	*Plegadis falcinellus*	Glossy Ibis	RM
62	*Pseudibis papillosa*	Red-naped/Black Ibis	R
63	*Threskiornis melanocephalus*	Black-headed/White Ibis	R
	Order: Coraciiformes		
	Family: Alcedinidae		
64	*Alcedo atthis*	Common Kingfisher	R
	Family: Cerylidae		
65	*Ceryle rudis*	Pied Kingfisher	R
	Family: Dacelonidae		
66	*Halcyon smyrnensis*	White-throated Kingfisher	R
	Order: Gruiformes		
	Family: Gruidae		
67	*Grus antigone*	Sarus Crane	R
68	*Grus grus*	Common Crane	M
69	*Grus virgo*	Demoiselle Crane	M
	Family: Rallidae		
70	*Amaurornis akool*	Brown Crake	R
71	*Amaurornis phoenicurus*	White-breasted Waterhen	R
72	*Fulica atra*	Common Coot	RM
73	*Gallicrex cinerea*	Watercock	R
74	*Gallinula chloropus*	Common Moorhen	RM
75	*Porphyrio porphyrio*	Purple Swamphen/Moorhen	R

Annexure 3. Habitat Requirement of Waterfowls in Thol Bird Sanctuary

R: Resident species, M: Migratory species, RM: Resident Migratory species

Habitat: L: Less used habitat, M: Moderately used habitat, H: Highly used habitat

8. References

Ali, Salim (2002). The book of Indian Birds. Oxford University Press, New Delhi.

Ambasht, R. S. (1990). A Text Book of Plant Ecology. Students' Friends & Co. Varanasi.

Anno. (1975) Gujarat State Gazetteers, Mehsana District. Government of Gujarat, Ahmedabad.

Anno. (2001) Management Plan of Thol Bird Sanctuary, Government of Gujarat, Gandhinagar.

Chase M. K., Kristan W. B. III, Lynam A. J., Price M. V. and Rotenberry J. T. (2000) Single species as indicator of species richness and composition in California coastal sage scrub birds and small mammals. Conservation Biology, 14(2) 474-487.

Croll, D. A. et al. (2005) Introduced predators transform subarctic islands from grassland to tundra. Science 307, 888-899

Deepa, R. S, and Ramachandra T. V., (1999) Impact of urbanization in the interconnectivity of wetlands, in proceedings of the National Symposium on Remote Sensing Applications for Natural Resources: Retrospectives and perspective, (Jan 19-21, 1999) organized by Indian Society of Remote Sensing, Bangalore, pp. 343-351.

FSI, (2011). State of Forest Report, Forest Survey of India, Dehradun.

Gaur, R. D. (1982) . Dynamics of vegetation of Garhwal Himalayas pp. 12-15. Vegetational Wealth of the Himalayas (ed. G. S. Paliwal) Delhi.

GEC, (2009). Preparation of Comprehensive Study Report Based on the Analysis of Various Ecological and Environmental Parameters to Implement the Action Plan for the Management and Conservation of Thol Wetland, Gujarat Ecology Commission, Gandhinagar.

GEER, (2002). Ecological study of Thol Lake Wildlife (Bird) Sanctuary, GEER Foundation, Indroda Park, Gandhinagar.

Grimmett, R., Inskipp, C. and Inskipp, T. (1998). Birds of the Indian Subcontinent. Oxford University Press, New Delhi.

Mc Farland, D. (1981). (Ed.) The Oxford Companion to Animal Behaviour, Oxford University Press, New York.

Mitsch, W. J. and Gosselink, J. G. (1986). Wetlands. New York: Van Nostrand Reinhold Comp. Inc., pp. 1-539.

Odum, E. P. (1983). Basic Ecology. Saunders College Publishing, Philadelphia and London.

Patel, Sejal & Dharaiya, Nishith (2008) Marsh Bird Community Index of Biotic Intigrity: A key to study an ecological condition of Wetlands.Proceedings of Tall 2007.

Patel, Sejal, Shingala P., Dave, S.M. and Dharaiya, N. (2006). A comparative study of avifaunal composition in two bird sanctuary of Gujarat. Proceeding of XXI Gujarat Science Congress, 2006.

Post, D.M. et al. (1998) The role of migratory waterfowl as nutrient vectors in a managed wetland. Conserve. Biol. 12, 910-920.

RAMSAR, (1971). Convention on Wetlands of International Importance, Ramsar, Iran.

Ravindranath, S. & Premnath, S. (1997) Biomass studies Field Methods for Monitoring Biomass. Oxford & IBH Publishing Co. Pvt. Ltd. New Delhi. pp 81

Singh, H. S. & Tatu, K. (2000) A study on Indian Sarus Crane (Grus antigon antigone) in Gujarat State (with emphasis on its status in Kheda and Ahmedabad Districts). GEER Foundation, Gandhinagar.

Singh, H. S. (1998). Wildlife of Gujarat (Wildlife and Protected Habitats of Gujarat State).GEER Foundation, Gandhinagar.

Singh, H. S., Patel, B. H., Parvez, R., Soni, V.C., Shah, N., Tatu, K. and Patel, D. (1999). Ecological study of Wild Ass Sanctuary, Little Rann of Kachchh (A Comprehensive study on Biodiversity and Management Issues). GEER Foundation, Gandhinagar.

Singh, R and Joshi, M. C. (1979). Floristic composition and life forms of sand dunes herbaceous vegetation near Pilani, Rajasthan. Indian J. Ecol. 6: 9-19

Singh, R. (1976). Structure and net community production of the herbaceous vegetation in the sand dune regions around Pilani, Rajasthan. Ph. D. Thesis BITS Pilani. 100p.

Vaghela, D. K. (1993) Fishes of Pond and Annual Survey. M. Phill. Dissertation, Department of Zoology, Gujarat University, Ahmedabad.

Vijayan, V. S. (1991). Keoladeo National Park Ecology Study. Final Report (1980-1990). BNHS.

Chapter 2

Biodiversity and Conservation Status of a Beech (*Fagus sylvatica*) Habitat at the Southern Edge of Species´Distribution

Rosario Tejera, María Victoria Núñez, Ana Hernando,
Javier Velázquez and Ana Pérez-Palomino

Additional information is available at the end of the chapter

1. Introduction

The aim of Habitats Directive (European Council, 1992) is "to contribute towards ensuring biodiversity through the conservation of natural habitats and of wild fauna and flora in the European territory of member states" (Article 2.1). This directive identifies a set of natural habitats and wild species of fauna and flora of Community interest (Annexes I and II of the Directive) and establishes the requirement to maintain a favourable conservation status. Therefore, Member States designate Special Protection Areas (SPAs), which are provisional sites of Community Importance (SCIs).

To ensure its enforcement, Member States should establish the necessary conservation measures involving, if necessary, appropriate management plans (Article 6.1).

According to the Article 1 of the Directive, the state of conservation of natural habitat is considered favourable when:

- its natural range and areas within that range are stable or increasing and
- the specific structure and functions necessary for long-term viability exist and are likely to continue to exist in the foreseeable future and
- the status of its typical species is favourable .

Member states have implemented different strategies for evaluating the conservation status of habitat types and species of Community interest, basing on both the European Commission reports (European Commission, 1995, European Commission, 2006, Shaw and Wind, 1997)) and scientific research (Bock et al., 2005, Dimitriou et al., 2006, Lang and Langanke, 2005, Noss, 1990, Noss, 1999, Roberts-Pichette, 1998, Simboura and Reizopoulou, 2007).

In addition, some Member States have developed their own methodologies for assessing the conservation status, such as Germany, Austria, the Netherlands, Portugal, Spain and the United Kingdom (Velázquez et al., 2010). These previous studies often propose numerical indicators and have been applied at regional or national levels (Cantarello, 2008).

In Spain, in 2009 the Ministry of Rural and Marine Environment issued a set of guidelines at national level to assess the conservation status of habitats and species of Community interest (AUCT. PL. , 2009). The main objectives of these guidelines are to identify and adequately describe the 117 habitat types and typify their conservation status.

2. Objetives

The objective of this study was to determine the conservation status of habitat 9120 - *Atlantic acidophilous beech forests with Ilex and sometimes also Taxus (Quercion robori-petraeae or Ilici-Fagenion)* within the beech forest of "Dehesa del Moncayo" (Spain) by applying the methodology provided by the Spanish Ministry. This implies a revision of the methodology at local level.

3. Methodology

The conservation status of habitats is assessed according to four general factors (European Commission, 2006): range and area occupied by the habitat, typical species, structure and function and future perspectives (Table 1). Each one can take the value of favourable, unfavourable-inadequate, unfavourable-bad or unknown. The overall assessment of the conservation status arises by combining the values obtained in Table 2 with the General Assessment Matrix (European Commission, 2006)

FACTOR	INDICATOR
Range and area occupied	Area (ha) and trend
Typical species	presence and abundance of typical species
Structure and function	Dead wood
	Forest structure
	Fragmentation
	Presence of Picidae
	Degree of defoliation
Future prospects	Current and potential threats

Table 1. Adaptation of the methodology for the habitat 9120.

PARAMETER	CONSERVATION STATUS			
	Favourable (green)	Unfavourable-inadecuate (amber)	Unfavourable-bad (red)	Unknown
Distribution area (range)	The range of habitat is stable (loss and expansion are balanced) or increasing and is not less than the "favourable area of reference"	Any situation other than those described in "green" or "red"	Large decrease in the range (equivalent to a loss of more than 1% per year over a period specified by the EC, other thresholds can be used but should be explained in Annex D Or the range is more than 10% below the "favourable reference range	Not available or insufficient reliable information
Area occupied by the habitat within the range	The area occupied by the habitat is stable (loss and expansion are balanced) or increasing and is not less than the "favourable area of reference " and without major changes in the distribution pattern within the range as a whole (if data are available for evaluation)	Any situation other than those described in "green" or "red"	Large decrease of the surface (equivalent to a loss of more than 1% per year over a period specified by the MS, other thresholds can be used but should be explained in Annex D Or with losses (negative changes) in the pattern of distribution within the range Or the current surface is more than 10% below the "favourable reference range"	Not available or insufficient reliable information
Structure and functions	Structures and functions (including typical species) in good condition and without significant damage/pressure	Any situation other than those described in "green" or "red"	More than 25% of the habitat is unfavourable in terms of its specific structures and functions (including typical species)	Not available or insufficient reliable information

Future prospects (regarding range, area covered and structure and function)	Future prospects are excellent / good, no significant effects of future threats, the long-term viability is guaranteed	Any situation other than those described in "green" or "red"	Future prospects are bad, serious impacts of threats, the long-term viability is not guaranteed	Not available or insufficient reliable information
Overall assessment of conservation status	All "green" or three "green" and one "unknown"	Any situation other than those described in "green" or "red"	Two or more "unknown" combined with "green" or all "unknown"	Not available or insufficient reliable information

Table 2. General Assessment Matrix

3.1. Distribution area and area occupied

The distribution area can be defined as "the current habitat areas " (AUCT. PL. , 2009). It aims to identify changes of distribution patterns of the habitat within the range. This factor makes sense at the biogeographic region scale. However, the range does not apply at the local level.

The area occupied assesses the area covered by the habitat in the study area and its trend:

1. Area occupied by habitat in the study area (in hectares).
2. Date of assessment.
3. Trend of area (stable, increasing, decreasing or unknown).
4. Magnitude of the trend.
5. Period of trend.
6. Reasons for the trend.

The concept of "Favourable Area of Reference" (FAR) shown on the General Assessment Matrix is defined as "the minimum area required within a biogeographic region to ensure long-term viability of a type of habitat" (European Commission, 2006). Neither this concept is of application for the current study, since it is a study at the local scale.

- Measuring procedure

For the present study, the vegetation map of the Moncayo Natural Park has been used. We have distinguished three main types of vegetation: beech (used for extracting charcoal), scot pine reforestated in the 19th century and natural Pyrenean oaks (Gallo Manrique, 2011).

- Assessment of conservation status

The conservation status was assessed based on the trend of the area occupied, giving a value of zero to the status in 2000, as proposed in the methodology.

3.2. Typical species

This factor considers the presence and viability of populations of typical species. That is, those that are indicators of habitat status. They can also be defined as those species relevant to maintain the habitat in a favourable conservation status, either because of their abundance or because of their influence in the ecological functions.

Typical species of the habitat 9120 are:

Flora: Yew (*Taxus baccata L.*), holly *(Ilex aquifolium L.), Lobaria pulmonaria L.*

Amphibians: Salamander (*Salamandra salamandra*)

Mammals: Gray dormouse (*Glis glis*)

Birds: White-backed woodpecker (*Dendrocopus leucotus*), Black Woodpecker (*Dryocopus martius)*, Nuthatch (*Sitta Europea*), Treecreeper (*Certhia familiaris*), Pied flycatcher (*Ficedula hypoleuca*), Marsh Tit (*Parus palustris*)

Invertebrates: saproxylic invertebrates: *Elona quimperiana, Rosalia alpina, Osmoderma eremita, Limoniscus violaceus, Cerambyx cerdo, Lucanus cervus, Gnorimus variabilis, Caliprobola speciosa.*

- Measuring procedure

The method used was based on observations of presence/absence of typical species during the field work reinforced with the wildlife catalog of Moncayo Natural Park (Gobierno de Aragón, 2002)

- Assessment of conservation status

It is not imperative that a particular location holds all or most of its typical species for a favourable conservation status (European Commission, 2006). But the set of all the habitats at the national or biogeographic scale must have long-term viable populations of all or many of the typical species of the habitat.

Since this study covers a small area of habitat 9120 in Spain, we assessed the number of typical species present in the forest. The result of this factor must be consistent with the structure and function factors.

3.3. Structure and function

Structure and function define the quality of habitat 9120 through four parameters: dead wood, stand structure, fragmentation, presence of *Picidae* and degree of of defoliation.

To determine the overall status of the structure and function, each indicator takes a value (0: unfavourable-bad, 1: unfavourable-inadequate, 2: favourable). The overall status of the structure and function can be unfavourable-bad — for results below 40% of maximum punctuation —, unfavourable-inadequate — from 40 to 75% —, and favourable — above 75%.

3.3.1. Dead Wood

This indicator measures dead wood (m³/ha), separating it according to: species, standing or fallen, size and level of decomposition.

-   Measuring procedure

The inventory of dead wood was done by strip-plots 500 m long and 20 m wide (1 ha), as proposed in the Spanish methodology (Olano and Peralta de Andrés, 2009). In these plots we measured dead wood — both standing and fallen —, diameter, length, species, and degree of decomposition.

The degree of decomposition was assessed according to the following criteria (Table 3).

Degree of decomposition	Description
Level 1	Healthy wood, with bark; wood intact
Level 2	Healthy wood, beginning of the bark loss
Level 3	Wood beginning to rot away. Without bark
Level 4	Very rotten wood, full of holes
Level 5	Completely rotten wood that breaks when touched

Table 3. Degree of d criteria

-   Assessment of conservation status

Dead wood in forests ranges from 10 to 150 m³/ha (Müller and Bütler, 2010). According to these authors, most species linked to dead wood seem to be present in hardwood forests for volumes between 30 and 50 m³/ha.

*   Unfavourable-Bad: less than 10 m³ of dead wood per hectare.
*   Unfavourable-inadequate: 10 to 30 m³ of dead wood per hectare, with at least 30% of deadwood above 30 cm diameter and 20% of standing dead wood.
*   Favourable: more than 30 m³ of dead wood per hectare, with at least 12 m³/ha of dead wood above 30 cm diameter and at least 4 m³/ha of standing dead wood. It is important that dead wood presents all stages of decomposition and it is distributed throughout the habitat.

3.3.2. Forest structure

Forest structure is evaluated according to three indicators: abundance of overmature trees (trees with dbh above 45 cm), structural diversity and species diversity. It is necessary to assess the number of stems/ha per diameter class and indicate the proportion of species.

To determine the overall status of forest structure, each indicator has a value (0: unfavourable-bad, 1: unfavourable-inadequate, 2: favourable). The overall status of the structure and function is unfavourable-bad — for results below 40% of the maximum punctuation —, unfavourable-inadequate — from 40 to 75% —, and favourable — above 75%.

- Measuring procedure

We inventoried diameter and species in circular plots 10 m radius, located at the points 100, 300 and 500 m of the strip-plots used for the inventory of dead wood.

- Assessment of conservation status

Overmature tree (dbh> 45 cm):

- Unfavourable-Bad: less than 5 trees/ha
- Unfavourable-inadequate: 6 to 10 trees/ha.
- Favourable: above 10 trees/ha

Species diversity:

- Unfavourable-Bad: less than 5 trees (dbh> 15 cm) /ha of other native tree species.
- Unfavourable-inadequate: 5 to 10 trees (dbh> 15 cm) /ha of other native tree species
- Favourable: above 10 trees (dbh> 15 cm) /ha of other native tree species

Structural diversity:

- Unfavourable-Bad: 90% trees in the same diameter class (classes of 20 cm).
- Unfavourable-inadequate: from 80% to 90% trees in the same diameter class (classes of 20 cm).
- Favourable: less than 80% trees in the same diameter class (classes of 20 cm).

3.3.3. Fragmentation

This indicator evaluates whether the habitat is a continuous patch of sufficient extent to ensure species survival or, conversely, is composed of individual patches.

Fragmentation is a very important element for forest communities that affects the quality of habitat and causes loss of species (Telleria and Santos, 2001). In beech forests, typical flora and fauna species are strongly affected by the edge effect due to their dependence on low light and high relative humidity.

- Measuring procedure

Fragmentation is quantified by comparing the total habitat area with the surface free of edge effect (effective area). We considered an edge effect of 30 m from the margins of the patches.

- Assessment of conservation status
  - Unfavourable-Bad: ratio between surface without edge effect and total area less than 80%.
  - Unfavourable-inadequate: ratio between surface without edge effect and total area from 80 to 90%
  - Favourable: ratio between surface without edge effect and total area above 90%.

3.3.4. Presence of Picidae

Picidae are known for tapping on tree trunks in order to find insects living in crevices in the bark and to excavate nest cavities. Some of these species require old forests with abundant dead wood, both standing and fallen. The presence of *Picidae* is a good indicator of habitat quality and conservation status.

- Measuring procedure

We performed a visual observation of cavities in the circular plots of the inventory. Additionally we used bibliographic survey (Gobierno de Aragón, 2002).

Assessment of conservation status
- Unfavourable-Bad: no *Picidae* nesting.
- Unfavourable-Inadequate: Only Great Spotted Woodpecker *(Dendrocopos major)* nesting.
- Pro: woodpecker White-backed Woodpecker *(Dendrocopos leucotos)* or Black Woodpecker *(Dryocopus martius)* nesting

3.3.5. Degree of defoliation

This indicator belongs to the group of indicators for the maintenance of health and vitality of forest ecosystems and is considered to be the main indicator of health status (MCPFE, 2002).

In Spain, there is a network of Forest Damage Assessment following the European methodology (International Cooperative Programme on Forests). In Moncayo Natural Park there are 5 plots for that network, but none of them within the beech forest "Dehesa del Moncayo".

- Measuring procedure

We visually assessed the percentage of defoliation in the circular inventory plots

- Assessment of conservation status

We used the thresholds of European Forest Damage Assessment Network (Table 4)

Defoliation class	% defoliation	Description
0	0-10%	No defoliation
1	>10-25%	Minimum
2	>25-60%	Moderate
3	>60-<100%	High
4	100%	Dead tree

Table 4. Criteria for assessing defoliation levels

3.4. Future perspectives

This factor refers to the long-term viability of a habitat considering possible threats, typical species and structure and function factors.

- Measuring procedure

We evaluated the main past and present impacts and the possible future threats that may affect the long term-viability of the habitat.

- Assessment of conservation status
  - Unfavourable: The future scenario does not ensure the long-term viability of habitat 9120
  - Favourable: The future scenario ensures the long-term viability of habitat 9120

4. Case study

4.1. Study area

The study area was the 1494 ha forest " Dehesa del Moncayo" within Moncayo Natural Park (Aragón, Spain) (Fig. 1). It is also included in the Natura 2000 Network as part of SCI ES2430028 "Moncayo" and SPA ES0000297 "Sierra del Moncayo-the-Fayos Sierra Arms" due to six habitats of interest. One of them is habitat 9120 *Atlantic acidophilous beech forests with Ilex and sometimes also Taxus in the shrublayer (Quercion robori-petraeae or Ilici-Fagenion).*

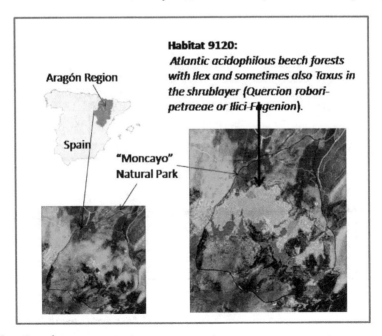

Figure 1. Location scheme

The forest is at the southern edge of the Mediterranean region. However, it is considered an "Atlantic island" due to the altitude and the NW-SE aspect. The Moncayo beech forest is between 1100 and 1900 m a.s.l. facing north or northeast with slopes between 20 and 50%. It has been mainly used for charcoal until 1940, since then it has evolved into a high polewood. It used to be also used for timber but, due to the poor quality of timber, cuttings have been infrequent. In 1978 it was declared Natural Park.

We distinguished four different forest types:

Beech forest (*Fagus sylvatica*)

This is a typical high density beech forest with holly (*Ilex aquifolium*) and blueberry(*Vaccinium myrtillus*) in less dense areas. It occupies 360 ha (Table 5)

Fagus sylvatica on screes

Small and branched isolated beech trees on rocky abrupt areas.

Rangeland of *Fagus sylvatica*

It is an area of small size (9.92 ha) used for grazing until 1920. As a result, big trees are accompanied by smaller trees.

Fagus sylvatica with heather (*Erica sp.*)

Beech forest with dense heather and other tree species such as Rebollo oak (*Quercus pyrenaica*) and Scots pine (*Pinus sylvestris*)

Forest types	Area	
	(ha)	(%)
Beech forest (*Fagus sylvatica*)	360,61	76,82
Fagus sylvatica on scree	94,76	20,19
Rangeland of *Fagus sylvatica*	9,92	2,11
Fagus sylvatica with heather (*Erica sp.*)	4,13	0,88
TOTAL	469, 42	-

Table 5. Forests types area (ha and %)

"*Fagus sylvatica* on scree" is assigned to habitat 8130 *Mediterranean and thermophilous screes*. Therefore it was excluded of the conservation status assessment.

4.2. Field survey

We performed a simple random pilot sampling inventory leaning on the network of paths. The pilot sampling was conducted over three consecutive days in July 2010.

The main objective of the pilot sampling was to calculate the variance and to determine whether the error was admissible or the inventory had to be strengthened with new sampling plots.

"Fagus sylvatica with heather" was excluded because it is a small area where the abundance of heather and the low density of trees do not justify the inventory.

4.3. Measuring procedure

4.3.1. Sampling units

We measured dead wood in four strip plots of 500 x 20 m and forest structure in 12 circular plots of 10 m radius (3 in each strip plot), distributed by forest type(Fig. 1).

It was decided to place them on the network of paths since this does not influence significantly the volume estimation of dead wood, due to narrow lanes (less than 1 m wide) and high bandwidth (10 m) both sides the path.

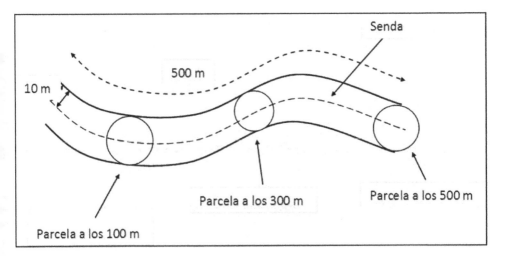

Figure 2. Sampling plots

We measured all dead wood on the ground from a minimum diameter of 10 cm (criterion given by the technical director of the study). Given the abundance of fine twigs on the ground and the large size of the sample plots (1 ha), measurement from 0 cm would have been impossible.

4.3.2. Measured variables

Variables from strip plots:

- Dead wood: Dead wood is classified into several groups: dead wood on the ground, standing dead trees, stumps and dead branches on living trees (Kirby et al., 1998). Diameter of the middle section (diameter at half the length of the fragment), length and level of decomposition was measured for dead wood on the floor. Diameter at 1.30m height, total height, and level of decomposition was assessed for standing dead wood.

Variables from circular plots:

- Forest structure: The reference methodology does not establish a minimum diameter for measuring forest structure. Following the technical director criterion, trees below 2.5cm diameter were excluded. Therefore, for the rest of the trees we measured all diameters at 1.30m height and recorded the species.
- Level of defoliation (by visual observation)
- number of cavities (natural or *Picidae*)
- number of trees below 2.5 cm diameter
- Mean height of the stand
- Description of the stand, indicating silvicultural characteristics, non target species, and a sketch/diagram/outline/schema of the vertical forest structure.

We performed a sheet for each plot.

4.4. Field work results

Analysis of variance (ANOVA) was performed for both variables to check significant differences between the two types of beech forest inventoried ("beech *Fagus sylvatica*" and "Rangeland of *Fagus sylvatica*"). The analysis showed no significant differences. So we adopted a single maximum admissible error for these variables.

When sampling dead wood, errors are generally quite high (Kirby et al., 1998, Van Wagner, 1982, Woodall et al., 2006, Woodall and Williams, 2005). Following Van Wagner (1982), in this study we assumed a 20% maximum admissible error.

Error for standing dead trees is higher than admissible (Table 8). However, lack of standing dead trees (Table 7) and heterogeneous distribution are typical of young beech forests.

Furthermore, the error for the variable basal area slightly exceeds the maximum so it was not considered necessary to reinforce the sampling.

strip plot	CIRCULAR PLOT	G (m²)	G (m²/ha)
	1	0,46	14,60
1	2	0,60	18,98
	3	0,75	23,88
	1	0,79	24,99
2	2	1,79	56,90
	3	0,91	29,06
	1	1,11	35,24
3	2	1,15	36,58
	3	0,96	30,48
	1	0,60	18,96
4	2	0,90	28,58
	3	0,61	19,53

Table 6. Basal area (G) by circular plot

	Dead wood volume (m³/ha)		
strip plot	total	Standing dead wood	dead wood on the floor
1	3,71	0,83	2,88
2	5,68	1,70	3,97
3	4,88	0,91	3,97
4	4,23	0,56	3,67

Table 7. Dead wood results by plot

VARIABLE	Mean	Variance	Error (%)
Basal area (m²/ha)	27,79	119,09	21,78
Dead wood on the floor (m³/ha)	3,62	0,27	14,26
Standing dead wood (m³/ha)	1,00	0,24	48,96
Total dead wood (m³/ha)	4,63	0,72	18,36

Table 8. Mean values, variance and error

5. Results for the conservation status assessment

5.1. Range

The results for the area occupied factor according to the methodology are:

Area covered by habitat 9120 within "Dehesa del Moncayo": 374.65 has
Date: 2011.
Trend: stable/increasing.
Trend-period: 1975-2011.
Reasons for the trend: the absence of human influence and good regeneration capacity.

5.2. Typical species

Typical 9120 habitat species present in "Dehesa del Moncayo" are the following:
Flora: Yew (*Taxus baccata*), holly (*Ilex aquifolium*) and *Lobaria pulmonaria*.
Holly is scarce except for some areas of low beech density. Yew and *Lobaria pulmonaria* are scarce or rare.
Birds: Nuthatch (*Sitta European*) and Pied flycatchers (*Ficedula hypoleuca*).
Invertebrates: *Rosalia alpina* and *Cerambyx cerdo*.
During the field work no typical species of fauna were inventoried.
Therefore, the conservation status for this factor is unfavourable-bad.

Typical species	UNFAVOURABLE-BAD

5.3. Structure and function

Snags are scarce and most of them are not large (Fig. 2)

Strip-plot	Volume of deadwood (m³m³/ha)		
	Total	Standing dead wood	Dead wood on the floor
1	3.71	0.83	2.88
2	5.68	1.70	3.97
3	4.88	0.91	3.97
4	4.23	0.56	3.67
Mean	**4.63**	**1.00**	**3.62**

Table 9. Volume of total deadwood — standing and on the floor — in each strip-plot

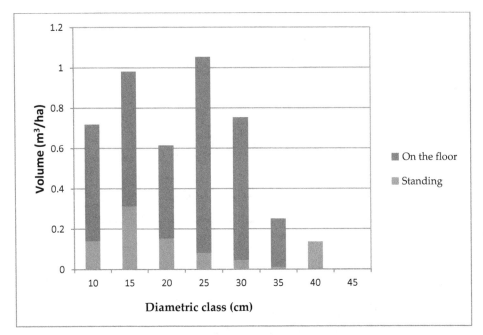

Figure 3. Volume (m³/ha) by diametric class and type of dead wood

The mean total volume of deadwood is below 10 m³/ha (Table 9). Therefore, the conservation status is unfavourable-bad.

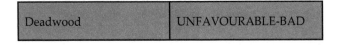

Deadwood	UNFAVOURABLE-BAD

5.3.1. Forests structure

- Oversized trees: Only 4 oversized trees were sampled (all of them in the "Rangeland of *Fagus sylvatica*" forest type) involving a total of 9.8 tree/ha. An unfavourable-inadequate conservation status was assessed for this component.
- Species diversity: Density of non target species with dbh above 15 cm was 14.7 tree/ha. Based on thresholds proposed in the methodology, a favourable conservation status for this component was assessed.
- Structural diversity:
- Total density reaches 1320 trees/ha. Above 50% of them are trees below 12.5 cm dbh. Regenerated beech (dbh< 2.5 cm)) is the most abundant diameter class, with 32.84% of total trees. Large trees are scarce.
- Less than 80% of trees group into the same diameter class (Fig. 3), so that the status regarding the structural diversity is favourable.

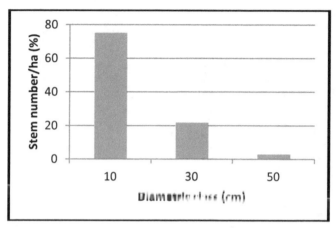

Figure 4. Tree distribution by diameter classes

Individual results (0: unfavourable-bad, 1: unfavourable-inadequate, 2: favourable) reach 5 of the 6 possible points. So the conservation status of forest structure indicator is favourable.

Forest structure	FAVOURABLE

5.3.2. Habitat fragmentation

Total area is 374.65 and the effective area, 278.36 ha. This yields a ratio of 74.30% free surface of the edge effect. According to the thresholds in the methodology, this is an unfavourable-bad status.

Fragmentation	UNFAVOURABLE-BAD

5.3.3. Presence of Picidae

The only species inventoried are Spotted woodpecker (*Dendrocopos major*), green woodpecker (*Picus viridis*) and Wryneck (*Jynx torquilla*) (experts information).

According to the thresholds established in the methodology, the conservation status is unfavourable-inadequate.

Picidae	UNFAVOURABLE - INADECUATE

5.3.4. Degree of defoliation

Defoliation reaches 12.5%. According to the pan-European forest monitoring criteria, defoliation level is low. Therefore, the conservation status is favourable.

DEFOLIATION	FAVOURABLE

5.3.5. Overall Conservation status of structure and function indicator

The overall conservation status scores 5 points (50% of maximum points), that means unfavourable-inadequate.

INDICATOR	Conservation status	Points
Dead wood	Unfavourable-bad	0
Forest structure	Favourable	2
Habitat fragmentation	Unfavourable-bad	0
Picidae	Unfavourable-inadecuate	1
Degree of defoliation	Favourable	2

Structure and function	UNFAVOURABLE -INADECUATE

5.4. Future Prospects

Given the low grazing pressure of herbivores, the control of public use and the low risk of fire, future prospects are favourable.

Future Prospects	FAVOURABLE

5.5. Global conservation status

Table 10 shows the results of the four general factors used to evaluate the conservation status of the habitat.

FACTOR	Conservation status	Global conservation status
Range	Favourable	**Unfavourable-bad**
Typical species	Unfavourable-bad	
Structure and function	Unfavourable-inadecuate	
Future Prospects	Favourable	

Table 10. Global Conservation Status for the 9120 habitat within "Dehesa del Moncayo" and final diagnosis

By applying the General Assessment Matrix (Table 2) criteria, we conclude that the conservation status of 9120 habitat within "Dehesa del Moncayo" is unfavourable-inadequate.

6. Discussion

The methodology of this study is an important step for assessing the conservation status of habitats of Community interest. The reference values are based on scientific research which should be adjusted periodically.

When applying this methodology to the 9120 habitat we found some difficulties:

The area of distribution and area occupied had to be adapted locally since the General Assessment Matrix proposes the biogeographic region. This led to only consider the area occupied by the habitat.

Measurement procedure for Typical species has not been standarised yet. We used bibliographic survey that may not accurately represent population and species of our habitat. The conservation status of typical species was unfavourable-bad due to the scarcity of species of fauna. However there may be several abundant and viable populations and more research would be necessary.

Structure and function is the core of the conservation status assessment. This Indicator was unfavourable-inadequate. Only forest structure and level of defoliation parameters had a favourable outcome. Although forest structure result was favourable, tree distribution is far from an uneven-aged forest which is the most suitable structure for biodiversity (Camprodon and Plana, 2007). Therefore, more research studies on thresholds of structure and function should be developed.

Procedures to measure dead wood have not been standarised yet either. Taking into account several studies (Kirby *et al.*, 1998, Woodall and Williams, 2005) we considered that "line transect" method could be more efficient than "strip-plot" method and could allow dead wood on the floor to be measured from 0 cm instead of 2.5 cm.

Finally, the overall conservation status unfavourable-inadequate shows the habitat is far from the favourable status. The lack of typical species of fauna is linked to the scarce dead wood and old trees with cavities. However, this is a young beech forest without productive exploitation since 1975 so the future prospects are favourable.

7. Conclusions

The Preliminary Ecological bases for the conservation of habitat types of Community interest in Spain (AUCT. PL., 2009) assesses the conservation status according to four general factors: range and area occupied, typical species, structure and function, and future prospects.

Although results showed an unfavourable conservation status, the current situation of the beech forest is the best one considering that it was highly harvested in the past. The future prospects are favourable and ensure the capacity of the forest to naturally achieve all the quality thresholds required, with no forest management actions.

Our results indicate that special attention must be paid to thresholds and that more accurate measurement procedures and assessment methods must be developed.

This methodology is an important and comprehensive starting point, however, it requires further applications to identify weaknesses and optimal measurement procedures.

Author details

Rosario Tejera
*Research Group for Sustainable Management, Department of Economy and Forest Management -
E.T.S.I de Ingenieros de Montes, Technical University of Madrid (U.P.M), Ciudad Universitaria,
Madrid, Spain*

María Victoria Núñez
*Research Group for Sustainable Management, Department of Projects and Rural Planning -
E.T.S.I de Ingenieros de Montes, Technical University of Madrid (U.P.M), Ciudad Universitaria,
Madrid, Spain*

Ana Hernando
*Research Group for Sustainable Management, Department of Economy and Forest Management -
E.T.S.I de Ingenieros de Montes, Technical University of Madrid (U.P.M), Ciudad Universitaria,
Madrid, Spain*

Javier Velázquez

Catholic University of Avila. C/ Los Canteros, Ávila, Spain

Ana Pérez-Palomino

E.T.S.I de Ingenieros de Montes, Technical University of Madrid (U.P.M), Ciudad Universitaria, Madrid, Spain

8. Acknowledgement

The research leading to these results has received funding from the Servicio Provincial de Medio Ambiente del Gobierno de Aragón and FEADER. We wish to thank all of the members of the research group "Silvanet" for their support and comments. We extend special thanks to Enrique Arrechea Veramendi for his cooperation and assistance and Miguel Valentín Gamazo for the English review.

9. References

AUCT. PL. (2009). *Preliminary ecological basis for conservation of habitat types of Community interest in Spain* Ministry of Rural and Marine Environment, Madrid.

Bock, M., Rossner, G., Wissen, M., Remm, K., Langanke, T., Lang, S., Klug, H., Blaschke, T. & Vrscaj, B. (2005). Spatial indicators for nature conservation from European to local scale. *Ecological indicators*, 5, 322-338.

Camprodon, J. & Plana, E. (2007). *Conservación de la biodiversidad, fauna vertebrada y gestión forestal*, Publicacions i Edicions Universitat de Barcelona.

Cantarello, E., Newton A.C. (2008). Identifying cost-effective indicators to assess the conservation status of forested habitats in Natura 2000 sites. *Forest Ecology and Management*, 256, 815-826.

Dimitriou, E., Karaouzas, I., Skoulikidis, N. & Zacharias, I. (2006). Assessing the environmental status of Mediterranean temporary ponds in Greece. *Annales De Limnologie-International Journal of Limnology*, 42, 33-41.

European Commission (1995). Standard Data Form for Special Protection Areas (SPA) for sites eligible for identification as Sites of Community Importance (SCI) and for Special Areas of Conservation (SCA).

European Commission (2006). Assessment, monitoring and reporting under Article 17 of the Habitats Directive: Explanatory Notes & Guidelines.

European Council (1992). Directiva 92/43/CEE del Consejo de 21 de mayo de 1992 relativa a la conservación de los hábitats naturales y de la fauna y flora silvestres. DO L 206 de 22.7.1992, p.7, Bruselas.

Gallo Manrique, P. (2011). *4ª Revisión del Proyecto de Ordenación del M.U.P. 251 "Dehesa del Moncayo", Tarazona (Zaragoza).*

Gobierno de Aragón, D.d.M.A. (2002). Plan Rector de Uso y Gestión del Parque Natural del Moncayo. Zaragoza.

Kirby, K.J., Reid, C.M., Thomas, R.C. & Goldsmith, F.B. (1998). Preliminary estimates of fallen dead wood and standing dead trees in managed and unmanaged forests in Britain. *Journal of Applied Ecology*, 35, 148-155.

Lang, S. & Langanke, T. (2005). Multiscale GIS tools for site management. *Journal for Nature Conservation*, 13, 185-196.

MCPFE (2002). Improved Pan-European Indicators for Sustainable Forest Management. Viena.

Müller, J. & Bütler, R. (2010). A review of habitat thresholds for dead wood: a baseline for management recommendations in European forests. *European Journal Forest Resource*, 129, 981-992.

Noss, R.F. (1990). Indicators for monitoring biodiversity - A hierarchichal approach. *Conservation Biology*, 4, 355-364.

Noss, R.F. (1999). Assessing and monitoring forest biodiversity: a suggested framework and indicators. *Forest Ecology and Management*, 115, 135-146.

Olano, J.M. & Peralta de Andrés, J. (2009). 9120 Hayedos acidófilos atlánticos con sotobosque de *Ilex* y a veces de *Taxus* (*Quercinion robori-petraeae* o *Ilici-Fagenion*). En: *Bases ecológicas preliminares para la conservación de los tipos de hábitat de interés comunitario en España*. (Ministerio de Medio Ambiente, y Medio Rural y Marino, Madrid.

Roberts-Pichette, P. (1998). Canada's Ecological Monitoring and Assessment Network with special reference to long-term biodiversity monitoring. *Forest Biodiversity in North, Central and South America, and the Caribbean*, 21, 47-56.

Shaw, P. & Wind, P. (1997). Monitoring the condition and biodiversity status of European Conservation Sites. Conservation, R.t.t.E.E.A.o.b.o.t.E.T.C.o.N.).

Simboura, N. & Reizopoulou, S. (2007). A comparative approach of assessing ecological status in two coastal areas of Eastern Mediterranean. *Ecological indicators*, 7, 455-468.

Telleria, J.L. & Santos, T. (2001). Fragmentación de hábitats forestales y sus consecuencias. En: *Ecosistemas Mediterráneos. Análisis funcional.* (Consejo Superior de Investigaciones Científicas y Asociación Española de Ecología Terrestre, Madrid. pp. 293-317.

Van Wagner, C.E. (1982). Practical aspects of the line intersect method. Canadian Forestry Service, Chalk River, Ontario, Canada.

Velázquez, J., Tejera, R., Hernando, A. & Nuñez, M.V. (2010). Environmental diagnosis: Integrating biodiversity conservation in management of Natura 2000 forest spaces. *Journal for Nature Conservation*, 18, 309-317.

Woodall, C.W., Rondeux, J., Verkerk, P.J. & Stahl, G. (2006). Estimating Dead Wood During National Forest Inventories: a Review of Inventory Methodologies and Suggestions for Harmonization En: *Proceedings of the Eighth Annual Forest Inventory and Analysis Symposium.*

Woodall, C.W. & Williams, M.S. (2005). Sampling Protocol, Estimation and Analysis Procedures for the Down Woody Materials Indicator of the FIA Program. United States Department of Agriculture (USDA), Forest Service.

Chapter 3

Top-Predators as Biodiversity Regulators: Contemporary Issues Affecting Knowledge and Management of Dingoes in Australia

Benjamin L. Allen, Peter J.S. Fleming, Matt Hayward, Lee R. Allen, Richard M. Engeman, Guy Ballard and Luke K-P. Leung

Additional information is available at the end of the chapter

1. Introduction

Large predators have an indispensable role in structuring food webs and maintaining ecological processes for the benefit of biodiversity at lower trophic levels. Such roles are widely evident in marine and terrestrial systems [1, 2]. Large predators can indirectly alleviate predation on smaller (and often threatened) fauna and promote vegetation growth by interacting strongly with sympatric carnivore and herbivore species (e.g. [3-5]). The local extinction of large predators can therefore have detrimental effects on biodiversity [6], and their subsequent restoration has been observed to produce positive biodiversity outcomes in many cases [7]. Perhaps the most well-known example of this is the restoration of gray wolves *Canis lupus* to the Greater Yellowstone Ecosystem of North America. Since the reintroduction of 66 wolves in 1995 [8], wolf numbers in the area have climbed to ~2000, some large herbivores and mesopredators have substantially declined, and some fauna and flora at lower trophic levels have increased (see [4], and references therein). Similar experiences with some other large predators mean that they are now considered to be of high conservation value in many parts of the world [1, 2, 7], and exploring their roles and functions has arguably been one of the most prominent fields of biodiversity conservation research in the last 10–15 years.

Large terrestrial predators are often top-predators (or apex predators), but not all top-predators are large or associated with biodiversity benefits [5, 9]. For example, feral cats *Felis catus* or black rats *Rattus rattus* may be the largest predators on some islands, but their effects on endemic fauna are seldom positive [10-13]. In geographically larger systems, coyotes (*Canis latrans*) [14] or dingoes (*Canis lupus dingo* and other free-roaming *Canis*) [15], for example, can exacerbate wildlife management problems in highly perturbed ecosystems, where they have the capacity to devastate populations of smaller prey [5, 16-18]. Hence, it is not the trophic position of a predator that determines their ecological effects, but rather their behaviour, impact and function [9]. This is most important for small- and medium-sized predators which can have positive, negative or neutral effects depending on a range of context-specific factors.

Excluding humans, dingoes are the largest terrestrial predator on mainland Australia but, at an average adult body weight of only 15–20 kg [19], are atypical top-predators [20-22]. No other continent has such a small top-predator, and canids have rarely (if ever) been a continent's largest predator, a role typically filled by ursids or felids. Australia's former terrestrial top-predator, a similar-sized marsupial known as the thylacine or Tasmanian Tiger *Thylacinus cynocephalus*, was quickly replaced by dingoes as the largest predator as thylacines became extinct coincident with the introduction of dingoes to Australia about 4000–5000 years ago [23-25]. Like all dogs, dingoes are derived from wolves by human selection [26-29], yet it is a mistake to equate dingoes with wolves (*sensu* [30, 31]) simply because they share a common origin [9, 22, 32] and display some wolf-like behaviours [19]. Hence, the net effects of dingoes on biodiversity might not be readily deduced from studies of other top-predators. Regardless of their derivation and exotic origin, dingoes are common across most of Australia's mainland biomes [33, 34], although their densities have been reduced to very low levels in some regions (<25% of Australia) where sheep *Ovis aries* and goats *Capra hircus* are farmed [15, 34].

Dingoes can have neutral, positive or negative effects (which can be either direct or indirect) on economic, environmental and social values [22, 35]. For example, dingoes can adversely affect livestock production by preying on livestock [36, 37], yet have beneficial effects to livestock producers by preying on livestock competitors [38, 39]. Alternatively, dingoes might help to reduce the impacts of smaller predators (such as introduced red foxes *Vulpes vulpes* or feral cats) on threatened fauna through intraguild predation or exploitative competition [40, 41], yet have detrimental effects on the same fauna through predation [15, 16] and/or disease transmission [42, 43]. Human attitudes towards dingoes are also variable [22, 44-46]. Hence, it should not be surprising to discover evidence for diverse and contrasting functions and values of dingoes in different places and at different times, which adds complexity to their best-practice management [35].

Knowledge of the roles of top-predators on other continents (e.g. [1, 2]) and recent research focus on the positive environmental effects of dingoes (e.g. [41, 47, 48]) has led to calls to cease lethal dingo control (e.g. [31, 49]) and even restore them to sheep and goat production regions (e.g. [23, 50]), actions collectively referred to hereafter as 'positive dingo management'. Serious concerns about the validity and rigour of the science supporting positive dingo management have been raised (e.g. [15, 51, 52], but see also [33, 53, 54]). The issue is further complicated by the changing genetic identity of dingoes [55-58] and the associated ambiguity and misuse of taxonomic terminology ([33]; e.g. compare taxonomic nomenclature between [56], [59], [60], and [55]). The capacity for dingoes to exploit seemingly unsusceptible fauna [61] and the widespread and direct negative effects of dingoes on biodiversity are also overlooked in many cases [15, 16]. There remains, however, a general view that dingoes provide net benefits to biodiversity at continental scales through suppression of foxes (Plate 1), feral cats and herbivores such as kangaroos (*Macropus* spp.) and rabbits (*Oryctolagus cuniculus*) [9, 47], and policy and practice recommendations towards positive dingo management are already occurring (e.g. [49, 62, 63]) despite concerns over the state of the literature and the conflicting roles of the dingo. In most places dingoes are presently managed on the basis of where they occur and what they are (or are perceived to be) doing, not on their genetics or appearance [33, 64].

Out of the confusion arise several knowledge gaps and issues which hamper the informed management of dingoes for biodiversity conservation. In this chapter we discuss critical knowledge gaps about dingo ecology, and highlight the influence of methodological application and design flaws on the reliability of published literature underpinning current knowledge of the ecological roles of dingoes. We offer alternative explanations for the mostly correlative data often mooted as 'clear and consistent evidence' (e.g. [54, 65]) for the fox-suppressive effects of dingoes, and discuss practical obstacles to the accrual of biodiversity benefits expected from positive dingo management. We also discuss the potential consequences of such a management approach for biodiversity and livestock industries, and the management of dingoes at scales which can address their context-specific impacts. Finally, we summarise some surmountable issues presently faced by researchers, land managers and policy makers, and provide recommendations for future research that, when completed, will assist in filling the knowledge gaps required to progress the best-practice management of dingoes for biodiversity conservation in Australia.

2. Knowledge gaps in the literature

Dingoes are one of the most studied animals in Australia, but there is still much to learn about them. Management of dingoes can be advanced by directing researchers towards critical knowledge gaps which require exploration. Unsurprisingly, some gaps need more urgent attention than others. Here, we focus on four key knowledge gaps that we consider to be fundamental to achieving best-practice management of dingoes as biodiversity conservation tools. These are:

1. The relationships between dingoes and biodiversity in relatively intact ecosystems
2. The relationships between dingoes and biodiversity in relatively altered ecosystems characterised by grossly disturbed vegetation structure and composition
3. The effects of current dingo control practices on mesopredators and biodiversity
4. The public's view of what we're trying to conserve (i.e. their pelage, their genetic identity and/or their ecological function)

Dingoes have been studied in many parts of Australia [19], but mostly in relatively intact (i.e. parks, reserves or extensive cattle production regions) and/or arid (Table 1) areas. This is mirrored by international research [2] that primarily comes from a limited number of classic studies conducted in relatively intact ecosystems that do not represent the majority of the earth's surface [66]. Although the relationships between dingoes and biodiversity in these intact areas might be considered well studied, they are not well understood, because the majority of the literature addressing the ecological roles of dingoes in these areas is compromised by a variety of methodological flaws [52]. Even ignoring these flaws, the majority of the relevant literature is only observational and correlative [41], and is therefore subject to plausible alternative explanations [67, 68]. Key among these is the cumulative effects of pastoralism (e.g. [15, 53]), which dramatically transformed pre-European landscapes into those characterised by severely altered vegetation communities [69-71] and a high proportion of now rare and locally extinct native fauna [72-75]. Understanding the roles of dingoes in highly altered ecosystems (i.e. sheep grazing lands and urban ecosystems) may actually be most important, because such systems are those expected to benefit most from positive dingo management [23, 50].

Since the 1960s, when the modern era of dingo research began, most studies have focussed on basic biology, including dingo diet, pack structure, physiology and reproductive biology [19, 76]. The motivation for much of this work has been directed at the negative effects of dingoes on livestock production [19, 64], and dingoes are presently subject to lethal control in many places in attempts to alleviate livestock predation [32, 64, 77]. However, due to the recently reported positive roles of dingoes and other top-predators on biodiversity conservation [1, 2, 7], lethal dingo control has come under increased scrutiny over its perceived indirect effects on biodiversity (e.g. [49]); the idea being that dingo control leads to negative outcomes for faunal biodiversity through trophic effects [23, 78]. Noteworthy however, is that the predicted negative effects of dingo *control* on faunal biodiversity are largely only presumed, and have rarely been demonstrated [79]. Regardless, the conservation and encouragement of dingoes is still being advocated on biodiversity conservation grounds (e.g. [23, 76]). However, *what* exactly requires conservation has not yet been determined for dingoes, which are listed as threatened species [56, 63] not because they are rare (in contrast, there are probably more dingoes now than at any other time in Australia's ecological history [33]), but because their genetic identity is again being altered through hybridisation [55, 57]. Unfortunately, phenotype or pelage is an unreliable indicator of genetic purity [58, 80], though most lay people equate purity with pelage (where only a sandy-coloured dingo is assumed to be pure). Alternatively, it may not be their colour or genetic identity that requires conservation, but their ecological roles [76]. Identifying what is to be conserved is important because most dingoes in Australia are not pure and are expected to become less so with time [55-57].

Understanding the trophic relationships between dingo management practices (i.e. poison baiting, trapping, shooting or no human intervention at all) and the conservation of threatened prey species (R1–R6 in Fig. 1) is the most critical management challenge [22, 41]. A wide variety of taxa may be involved (Plate 1). Ecological relationships between organisms are rarely as simple as those described in Fig. 1, yet they are often assumed to be so in studies of dingoes [32]. The (mostly negative) relationships between exotic mesopredators and threatened prey species (R3) are relatively well understood from other studies [81, 82], as is the relationship between lethal dingo control and dingoes (R1) [64, 83]. The other two relationships (R4 and R6) have received less attention (Table 1), although these are arguably the two relationships most able to address questions relating to the trophic consequences of dingo control. The direct risks dingoes pose to threatened fauna (R5) should also be well established before positive dingo management can be implemented with confidence [22]. Dingoes are highly adaptable and generalist predators capable of threatening many of the species they have also been predicted to protect [16, 17]. Studies that focus on R2 (and report that dingoes are negatively associated with foxes and cats) typically presume that lethal control of dingoes must therefore benefit foxes and cats (R4), though such an assumption is unfounded [22, 32]. Of ultimate importance however, and irrespective of any of the other relationships, understanding the effect of dingo control on threatened prey species (R6) can facilitate the most rapid management progress. The short-term and direct effects of dingo control on threatened fauna were reviewed in [79], which concluded that no studies to date have shown negative effects of dingo control on non-target fauna, a view subsequently ratified in [84]. There remains, however, limited reliable data on the longer term and indirect effects of dingo control faunal biodiversity [41, 85].

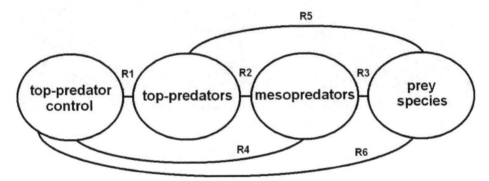

Figure 1. Schematic representation of six relationships (R1–R6) between top-predator control and prey species at lower trophic levels.

Investigating R6 is a 'black box' approach to applied research [86], meaning the observed outcomes of control interventions can enable management progress in the absence of a complete understanding of the mechanisms responsible for the outcomes. For example, [86] summarised the results of 25 years of experimental research on the conservation of threatened black-footed rock-wallabies *Petrogale lateralis*, stating that researchers had found time and again that fox control resulted in more rock-wallabies, but they did not have a good grasp on the mechanisms responsible for it. Thus, if investigations of R6 show that threatened prey populations fluctuate independently of dingo control, lethal control of dingoes might continue to occur without concern from conservationists that such practices inhibit the recovery of threatened fauna through trophic effects. Lethal dingo control may not be incompatible with biodiversity conservation or restoration [32], nor is cattle production always incompatible with dingoes in the absence of dingo control [38, 87, 88]. In a world where resources to manage threatened species are limited, focussing on such applied studies should be of utmost value to land managers and policy makers.

Plate 1. Rufous hare-wallabies *Lagorchestes hirsutus* (bottom right; photo from www.arkive.org), dusky hopping-mice *Notomys fuscus* (bottom left; photo by Reece Pedler) and red foxes *Vulpes vulpes* (top right; photo by Ben Allen) are some of the fauna that are affected both positively and negatively by dingoes (top left; photo by Ben Allen).

3. The state of current evidence for dingoes' ecological roles

Classical manipulative experiments are the best way to advance scientific knowledge [89, 90]. However, performing robust experiments on dingoes at large-enough scales is costly and logistically very difficult or even impossible [41]. Almost all field studies typically sample dingo populations using passive tracking indices (or sand plots) placed along dirt roads and trails. The use of other monitoring techniques, such as camera trapping, are increasingly being used [91, 92]. Although many studies investigating R2 and R5 using passive tracking indices have claimed to provide evidence that dingoes stabilise ecological processes through their top-down effects on sympatric predators and prey, three unresolved issues continue to compromise the reliability of these conclusions for most studies (Table 1):

1.　Much of the literature is weakened by methodological flaws (such as seasonal or habitat confounding, or invalid and violated assumptions) which render the reliability of the body of data collected uncertain [52]. In many cases, it is not the technique that is weak, but it is the poor application of otherwise robust techniques that compromise the data collected [51]. This is not to say that the conclusions of such studies are incorrect, but that the reader cannot tell whether they are or not because of the flaws.

2. Regardless of their methodological flaws, most studies are also conducted over small
    spatial and/or temporal scales. Because of spatiotemporal variation in animal densities
    [67, 93, 94], behavioural avoidance of top-predators by mesopredators [3, 95, 96], and
    because most studies sample dingoes along roads (which are favoured by dingoes;
    [95]), the results of many recent studies may simply be artefacts of sampling biases
    towards apparent inverse relationships between dingoes and mesopredators.
3. Regardless of methodological flaws or sampling bias, the experimental designs of many
    studies are still only observational or correlative ([41]), rendering their conclusions
    subject to a wide variety of plausible alternative explanations [53, 68]. Such studies can
    only support statements such as 'dingoes *might* perform this role' instead of statements
    such as 'dingoes *do* perform this role', which can only be made reliably from studies
    with greater inferential capacity [89].

3.1. Methodological flaws

Critical review has shown that the data in 75% (15 of 20) of recent studies that sampled
dingoes using sand plots on roads are potentially confounded by a variety of factors,
including (but not limited to) invalid seasonal and habitat comparisons [52]. Dingo activity
on roads varies between seasons independent of their actual abundance [52, 97], which can
lead to confounding and weakened inferences if not accounted for by the study design. For
example, valid comparisons cannot be made between one site sampled in winter and
another site sampled in summer, because observed activity differences are likely to be
attributable to behavioural changes and not abundance changes. This issue may most easily
be understood for reptiles, which usually reduce their activity in winter [98]. For dingoes
and foxes, food availability and breeding may drive this variability [19, 99].

Comparisons between different habitats may also be confounded due to varying detection
probabilities associated with different habitat types [68, 93]. For example, even if abundance
is equal across habitats, animals occupying landscapes with more difficult terrain may
utilise roads (i.e. where sampling occurs) more frequently than animals occupying areas
which allow more ubiquitous movements (e.g. [100]), with observed activity differences
again potentially attributable to behavioural changes and not abundance changes.
Moreover, different habitats often have different faunal assemblages, geological and
ecological processes (e.g. [101]), which may influence the way some species interact with
sand plots placed on roads. Pooling across seasons or habitats may mask differences that
could be more easily viewed if separated (e.g. [32]). A variety of assumptions (such as
'footprints of the same species <500m apart and heading in the same direction belong to the
same individual' or 'old-looking footprints are *x* days old') are also commonly made (Table
1) and undoubtedly violated ([52]; but see [88, 102-104] for examples). Violation of such
assumptions may underestimate dingo distribution or abundance.

Although a wide variety of methodological flaws are evident (Table 1), violation of
assumptions and seasonal or habitat confounding may be more important than other flaws,
in that they could have greater ecological significance than other methodological errors [52,

93]. Of the 34 studies considered in Table 1, 14 (41%) and 15 (44%) and are potentially weakened by habitat and seasonal confounding, while 12 (35%) made unnecessary assumptions, indicating that multiple studies contain multiple methodological weaknesses. Fundamentally, indices are only useful when they are correlative of abundance [67, 105], and such flaws typically mean that the relationship between observed indices and actual abundances is unknowable. We note however, that accurate knowledge of absolute abundance is near impossible to acquire in the field [67, 105, 106], and we are not aware of any studies of dingoes that have calibrated sand plot activity data with absolute abundance values (because absolute abundance values have not been attainable). However, where the principles outlined in [93, 106] are strictly applied, researchers can acquire reliable estimates of relative abundance, the metric that underpins the vast majority of available field data on dingoes (Table 1).

The use of inappropriate techniques or poor application of otherwise robust techniques reduces the extent to which such data can be used to make reliable statements about ecological processes, and because many studies have made such flaws (Table 1; [52]), much of the available sand plot data on dingoes might be considered unreliable. Overturning this conclusion for any given study requires demonstration that either (1) the methodological flaws described were not made and/or (2) that if made, they did not constitute unreliability [53]. Once collected, it is also rarely possible to un-confound the data using statistical procedures (such as generalised linear modelling) without making the most tenuous of assumptions [52, 105]. The design flaws outlined here are discussed in more detail in [33, 52]. Others [53, 54] have questioned the importance of these flaws, but such methodological flaws are not the only issue undermining evidence for dingoes' ecological roles.

3.2. Sampling bias

An index is a measurement related to the actual variable in question [67, 105, 107] and specific to the circumstances under which the data were collected [93]. Importantly, animal populations are not usually distributed uniformly across the landscape but are instead clumped, producing areas of higher and lower abundance (e.g. [108]). Thus, studies conducted over small spatial scales may acquire severely biased results. For example, the areas sampled in [109] or [110] were very small (<10km2), which likely represented only a fraction of a dingo's home range in such systems [111, 112]. The observed relationships between species within such small areas may have limited applicability outside the areas sampled, where animal abundances may be markedly different (e.g. [108]). Animal activity is also rarely distributed uniformly over temporal scales. Within a 24 hour period, animals may exhibit diurnal, nocturnal or crepuscular behavioural cycles which prevent reliable comparisons of index values from one time period to another. This may be most easily understood for birds, where, for example, observations collected from one area in the early morning should not be compared to observations collected from another area at noon [113, 114]. Many of these considerations essentially amount to issues of detection probability, and have been discussed in greater detail elsewhere [68, 93, 114, 115]. The same principles apply to indexing and population estimation using almost any technique [93, 116].

Top-Predators as Biodiversity Regulators: Contemporary Issues Affecting Knowledge and Management of Dingoes in Australia

65

The highest activity periods for top-predators are also usually optimal, mesopredators usually avoid top-predators during these times, and prey activity usually fluctuates independently of predator activity (e.g. [117-119]). Because mesopredators typically seek to avoid encountering top-predators, mesopredator activity is likely to be lower at times and in places with higher top-predator activity. This has important implications for studies conducted over restricted temporal scales, such as snap-shot or single sample studies (Table 1; e.g. [120-122]). If dingo activity is high on those days, mesopredator activity would be expectedly lower (and vice versa), which means that such temporally limited data is silent on the ability of dingoes to suppress or exclude mesopredator abundances over time, because mesopredators may simply have been avoiding the sampling area on those days. Repeating this snap-shot approach to sampling at any number of multiple sites cannot overcome this issue of bias. Conducting successive surveys over slightly longer timeframes (e.g. three or four surveys over one year) may also be affected by this bias because periods of high or low top-predator activity may endure for several months [52, 97, 111, 123]. Some such studies (e.g. [110, 124]) might been viewed as positive population responses of mesopredators to single dingo control events. Again, however, such observations would be expected given that mesopredator behaviour may change, increasing their use of tracks once the landscape of fear has been altered [96, 125, 126] without necessarily altering their actual abundance (e.g. [110, 124, 127]). Temporally restricted data cannot be reliably used as evidence that dingo control increases the abundance of mesopredators unless the results can be adjusted for seasonal effects by incorporating data from a comparable nil-treatment area. Even over several years, a sampling strategy which focuses on landscape features where dingoes are expected to be more active (such as dirt roads and trails) are also likely to be biased towards dingoes and less sensitive (but not insensitive; e.g. [87]) at detecting foxes or cats [95].

Such issues of bias on sand plots are typically overcome by sampling populations over larger spatial and/or temporal timeframes [93] and means that interspecific comparisons of index values are inappropriate [93, 94]. Other population sampling and analytical techniques might be used (such as estimates derived using photo-mark-recapture [128-131], camera trap rates [132], aerial surveys [133, 134], distance sampling of actual observations or signs [113], occupancy modelling [68] or track transects [135]), but these are all likewise subject to similar issues [114, 116]. Even though magnitudes of index values are meaningless for comparison between species, the population trends defined by the index values over time can be valid given appropriate study design and data analyses [93]. All studies identified in Table 1 have sampled predators for only a few days at a time during each survey, meaning that the results from each individual survey, in isolation, might be artefacts of such bias. This is an important weakness of short-term studies, but when surveys are repeated over several seasons or years, resulting trends may be reliably used to identify relationships between predators. For example, fox activity on sand plots may be much lower than those of dingoes for any (or every) given survey (possibly as a result of sampling bias), but when surveyed repeatedly over longer timeframes, correlations between dingo and fox population trends can be confidently compared. When dingo abundance is further manipulated in an experimental framework, a divergence of activity (or relative abundance) trends between dingoes and foxes would be particularly strong evidence for mesopredator suppression or release. The corollary of this is that non-divergence of dingo and fox population trends over time would be particularly strong evidence that mesopredator suppression by dingoes is not occurring.

Additional to the methodological flaws described earlier, many studies are also conducted over small spatial or temporal scales (Table 1). Thus, their results are likely to be affected by the sampling biases described, giving the potentially mistaken impression of inverse relationships between dingoes and mesopredators. The common presence of this issue throughout the literature further weakens the reliability of data on dingoes' ecological roles. Such biased data might only be suggestive of spatial avoidance between predators, but it cannot demonstrate avoidance. Provided the proper indexing principles are strictly applied and the data analysed appropriately, studies assessing predator population trends over longer timeframes will have a much better ability to identify correlative relationships. However, to identify causal process for observed correlations still requires experimental designs with even greater inferential ability [89, 90].

3.3. Experimental design

Poor application of methods and sampling bias are but two forms of experimental design flaws weakening the reliability of many studies. But even if such issues are overcome through appropriate sampling strategies, different types of experimental designs have inherent limitations to their inferential ability [89]. The implications of these limitations have not been adequately dealt with in most appraisals of the literature on dingoes' ecological roles. In 2007, [41] concluded that the available data on dingoes' ecological roles was mostly observational and correlative, and many studies published since then (e.g. [31, 78, 122, 136-138]) have not improved this situation. It should be understood that 'studies of a more observational nature can make only weak inferences about cause and effect and studies that involve classical experiments can make stronger inferences. Where studies use more observational methods the results should be interpreted and valued as such, and not as equivalent to the results of classical experiments' ([89]; but see also [90]). The replication and randomisation of treatments, along with the use of nil-treatments (or experimental controls) are particularly important design features that can provide a greater ability to demonstrate causal processes – provided methodological flaws and sampling bias are also avoided.

The inferential capabilities of different designs used in 34 studies of dingoes are here ranked between 1 and 16 (1 = highest level of inference, 16 = lowest; from [89]) in Table 1. Without a nil-treatment, the highest rank a study can achieve is a pseudo-experiment type I (Rank 9). Without randomisation, the highest rank possible is a quasi-experiment type I (Rank 5). For studies comparing the effect of contemporary or historical dingo control practices on predators or prey, many researchers cannot randomise their treatments and are constrained to use areas where dingo control is (or is not) already being undertaken (e.g. [83, 139]). In the case of cross-fence comparisons (e.g. [78, 122, 140]), the results of such non-randomised studies may be subject to plausible alternative explanations that cannot be controlled for [15,

Top-Predators as Biodiversity Regulators: Contemporary Issues Affecting Knowledge and Management of
Dingoes in Australia

67

101, 121]. Where possible, treatment randomisation offers one way of addressing these constraints, but has only been undertaken by three studies (Table 1). Only one study [32] has involved a classical experiment on dingoes, where treatments and nil-treatments were also replicated (two of each at one site). Thus, almost all of the available literature reports results from experimental designs which cannot reliably demonstrate cause and effect. Each of these three issues (methodological flaws, sampling bias and experimental design limitations) mean that the evidence for dingoes' ecological roles is not as strong as might be supposed, and each of these issues must be overcome in order to change this view.

As an example of how these issues combine to effect the reliability of data, [121] used footprint counts on dirt roads to derive activity indices for dingoes, foxes and cats at three sites on either side of the dingo barrier fence, which was erected in the early 20th century to exclude dingoes from sheep production lands in south-eastern Australia [141-143]). At two sites, fox activity was reportedly ~2–3 times higher in places where dingoes were rare. At a third site, foxes were only detected where dingoes were rare, and cats were reportedly present in equally low abundance on both sides of the fence [121, 138]. The methodological flaws described earlier (and in [52]) mean that the results of [121] could only be considered 'coarse measures'. Although, [53] argued that coarse measures are sufficient in places where the effect sizes are too large to be explained by the methodological shortcomings (such as seasonal confounding), meaning that the quantitative data may be unreliable but the qualitative patterns may still be recognisable. Importantly however, predator activity can naturally vary in excess of 400% in a matter of weeks or months (e.g. [32, 83, 144]), which means that the effect sizes must be enormous for comparisons made between different seasons to not be affected by season. Regardless, sampling occurred only once over a few days at each of the three sites described in [121]. Because, in such habitats, mesopredators typically avoid roads and dingoes do not [95], the low incidence of fox tracks in the presence of greater numbers of dingo tracks could simply be an artefact of spatial avoidance of roads by foxes on the days that footprint counts were collected. This result may not necessarily reflect the relative abundance of foxes at all, because foxes may have been more active in other parts of the landscape on those days – the infrequent detection of mesopredator tracks would be expected at a time of high top-predator activity (or vice versa). Whether the methodological flaws or the potential for sampling bias are considered important or not, [121] was still only a non-randomised correlative quasi-experiment type I [89], with an inferential rank of 5 out of 16 (Table 1). Hence, the observations may equally be explained by alternative factors, such as the cumulative impacts of livestock grazing [15, 121], thus offering only 'inconclusive' support [53] for the functional relationships between the species studied.

We are not trying to argue here that foxes are actually abundant on the same side of the fence as high-density populations of dingoes, or that dingoes are actually abundant on the same side of the fence as high-density populations of foxes. Rather, we seek only to illustrate that the sampling biases inherent to short-term studies prohibit the demonstration of causal relationships. In no way is the preceding discussion on the state of the literature intended to be personally critical of researchers and authors, because achieving robust experiments is logistically very difficult [41] and randomisation of treatments is often impossible. Rather, we simply aim to show that whether it is methodological flaws or sampling bias or experimental design limitations, most studies cannot provide strong evidence for causal factors associated with dingoes' ecological roles. It is also important to remember that because perfect experimental designs can be executed imperfectly and imperfect designs may be executed perfectly, neither may enable reliable inference. In other words, correlative or mensurative studies that avoid the flaws and biases described may be just as inconclusive as experimental studies that contain them. As [145] cautioned, 'don't even start the project if you cant do it right', because if the basics are not right, such projects may 'only represent wasted resources' [115].

Reference	Study topic (climate)	Methodological strengths	Methodological weaknesses	Spatial scale per site & sampling effort	Relation-ships investigated^	Experimental design (highest rank of inference)*
Allen B.L. [32]	The effect of dingo control on dingoes (arid)	Manipulative experiment BACI design Random allocation of treatments Treatment replication at some sites Time-series data	Baiting intensity varied within treatments between replicates	50 plots over 50km (x2) 6–10 counts at 4 sites over 2–4yrs	R1	Classical experiment (1) & Unreplicated experiment (3)
Allen L.R. [87]	The effect of dingo control on beef cattle (monsoonal tropics and semi-arid)	Manipulative experiment BACI design Random allocation of treatments Time-series data	No replication at individual sites	50 plots over 50km (x2) 7–19 counts at 3 sites over 3–4yrs	R1, R4, R5, R6	Unreplicated experiment (3)
Allen L.R. [83]	The effectiveness of dingo control campaigns (semi-arid)	Replication of treatments Multiple properties surveyed Temporally intensive sampling Time-series data	Non-random allocation of treatments Non-independence between treatments Baiting intensity varied between properties within-treatments	92–133 plots over 92–133km 16–23 counts at 3 sites over 2–3yrs	R1	Quasi-experiment type I (5)
Augusteyn et al. [146]	The effect of dingo control on dingoes and bridled nailtail wallabies	BACI design Manipulative experiment Time-series data Measured demographic responses of prey	One study site only No nil-treatment	53 plots over 53km 20 counts at 1 site over 5yrs	R1, R2, R5, R6	Pseudo-experiment type VII (15)
Brawata & Neeman [140]	Predator distribution around waterpoints in the arid zone (arid)	Spatial replication of treatments Two indices of predators used	Data confounded by habitat and seasonal effects Used binary observations over potentially continuous measures Two experiments in one, but analysed together Sand plot index data untransformed	15 plots over 20km (x2) and 20 scent stations over 20km (x2) 2 counts at 5 sites over 3yrs	R1, R2, R4	Quasi-experiment type I (5)

Reference	Study topic (climate)	Methodological strengths	Methodological weaknesses	Spatial scale per site & sampling effort	Relation-ships investigated^	Experimental design (highest rank of inference)*
Burrows et al. [147]	The effects of dingo control on dingoes, foxes and cats (arid)	BACI design Three indices of predators attempted Time-series data	Non-random allocation of treatments Invalid assumptions when calculating the activity of predators Data confounded by seasonal differences in predator activity Invalid comparisons between species One index technique (cyanide bait uptake) removed individuals from the population	30–60km tracking transects 25 counts at 1 site over 10yrs	R1, R4	Quasi-experiment type III (7)
Catling & Burt [148]	The influence of habitat on small mammals (temperate)	Mensurative study Standardised design	Data confounded by seasonal differences in predator activity Invalid comparisons between habitats Sand plot index data untransformed	20–35 plots over 4–7km 2 counts at 13 sites over 7yrs	R3, R5	Pseudo-experiment type V (13)
Catling et al. [149]	The effects of cane toads on native fauna (monsoonal tropics)	BACI design Three treatments Different indices for some species	Used binary observations over potentially continuous measures Sand plot index data untransformed	25 plots over 5km 4 counts at 1 site over 2yrs	R5	Quasi-experiment type I (5)
Christensen & Burrows [150] (see also [147])	Reintroduction success of native mammals following predator control (arid)	Two measures of predators used	Invalid assumptions when calculating the activity of predators Predators in 'nil-treatment' areas sampled using an index technique (lethal cyanide bait uptake) that removed individuals from the population 'Nil-treatment' area relocated during the course of the study Cyanide sampling technique biased towards dingoes and foxes Only 1 (of 2) treatment was sampled on 7 of the 8 surveys Not all survey results are reported No analyses undertaken	60km tracking transect 8 surveys at 1 site over 4yrs	R1, R2, R3, R4, R5, R6	Quasi-experiment type IV (8)
Claridge et al. [151]	The effect of predator control on activity trends of forest	Mensurative study Spatial replication of treatments and transects Time-series data	Used binary observations over potentially continuous measures Assumed independence between sand plots	75-125 plots over 19-31km 19 counts at 1 site over 9yrs	R1, R4, R6	Quasi-experiment type I (5)

Reference	Study topic (climate)	Methodological strengths	Methodological weaknesses	Spatial scale per site & sampling effort	Relation-ships investigated^	Experimental design (highest rank of inference)*
	vertebrates (temperate)					
Corbett [152]	Relationships between dingoes, water buffalo and feral pigs (monsoonal tropics)	BACI design Independent indices of some species Calibrated pig and dingo indices with mark-recapture estimates and total counts Time-series data	Used binary observations over potentially continuous measures	55 plots over 400km 27 counts at 1 site over 7 yrs	R5	Quasi-experiment type I (5)
Edwards et al. [102]	Habitat selection by dingoes and cats (arid)	Mensurative study Standardised design	Invalid assumptions when calculating the activity of predators Data confounded by seasonal and habitat differences in predator activity	25km tracking transects (x4) 9 counts at 1 site over 3yrs	R2	Psuedo-experiment type V (13)
Edwards et al. [153]	The effect of rabbit warren ripping on wildlife (arid)	Spatial replication of treatments	Invalid assumptions when calculating the activity of predators Data confounded by seasonal and habitat differences in predator activity Baiting intensity varied between sites	10km tracking rectangle (x2) 8 counts at 4 sites over 2yrs	R1, R2, R5	Quasi-experiment type I (5)
Edwards et al. [154]	The effect of Rabbit Haemorrhagic Disease on wildlife (arid)	Mensurative study Standardised design	Invalid assumptions when calculating the activity of predators Data confounded by seasonal and habitat differences in predator activity Data influenced by rabbit warren ripping at some sites	10km tracking rectangle (x2 at four sites) 8 counts at 6 sites over 2 yrs	R2, R3, R5, R6	Pseudo-experiment type V (13)
Eldridge et al. [88]	The effect of dingo control on dingoes and wildlife (arid)	Manipulative experiment Two measures of predators used	Invalid assumptions when calculating the activity of predators	10km tracking transects (x6) 7 counts at 3 sites over 3yrs	R1, R4, R6	Unreplicated experiment (3)
Fillios et al. [155]	Relationships between dingoes and kangaroos (arid)	Spatial replication of treatments Independent measures of kangaroos and dingoes	Replication devalued by seasonally staggered indexing Data confounded by seasonal and habitat differences in predator activity	25 plots over 25km (x2) 1 count at 6 sites over 1yr	R5	Quasi-experiment type I (5)

Top-Predators as Biodiversity Regulators: Contemporary Issues Affecting Knowledge and Management of
Dingoes in Australia

71

Reference	Study topic (climate)	Methodological strengths	Methodological weaknesses	Spatial scale per site & sampling effort	Relation-ships investigated^	Experimental design (highest rank of inference)*
			Sand plot index data untransformed			
Fleming et al [139] (see also [156])	The effects of dingo control on dingoes (temperate)	BACI design Index data transformed Data corrected for detection probability	Non-random allocation of treatments Abundance and activity potentially confounded	120–270 plots over 12–27km (x2) 12 counts at 1 site over 3yrs	R1	Quasi-experiment type 1 (5)
Johnson & VanDerWal [136] (using data from [157, 158])	Dingoes ability to limit fox abundance (temperate)	Source data from mensurative studies Large data set over wide spatial distribution	Source data confounded by seasonal and habitat differences in predator activity Source data used binary observations over potentially continuous measures Invalid comparisons between species Sand plot index data untransformed	From [158]: 45 plots over 18km, 65 plots over 26km and 105 plots over 84km Repeated counts at 3 sites for up to 9yrs From [157]: 20–35 plots over 4–7km 1 or 2 counts at 15 sites over 7yrs	R2	Pseudo-experiment type V (13)
Kennedy et al. [159]	Relationships between dingo control, dingoes and cats (monsoonal tropics)	Mensurative studies and manipulative experiments Spatial replication of treatments Mensurative study temporally replicated Data transformed Time-series data	Site differences not explicitly identified Temporal trends in predator activity not reported	30–50 plots over 30–50km (x10) 3 counts at 2 sites over 3 years, 2 counts at 2 sites over 2–4 weeks	R1, R2, R4	Pseudo-experiment type I (9) & Quasi-experiment type 1 (5)
Koertner & Watson [160]	The impact of dingo control on quolls (temperate)	Uses two measures of efficacy Replication of treatment (individuals exposed)	Used binary observations over potentially continuous measures Index data untransformed	36 plots over 36km 2 counts at 1 site once	R1, R4	Quasi-experiment type I (5) & Pseudo-experiment type V (13)
Letnic et al. [121] (a subset of [122])	Dingoes' role in protecting dusky hopping-mice from predation by foxes and cats (arid)	Spatial replication of treatments Different measures for hopping-mice and dingoes	Replication devalued through seasonally staggered indexing Insensitive measures of grazing pressure used Data influenced by seasonal and habitat differences in predator activity	25–30 plots over 25–30km (x2) 1 count at 3 sites over 1yr	R3, R5	Quasi-experiment type I (5)

Reference	Study topic (climate)	Methodological strengths	Methodological weaknesses	Spatial scale per site & sampling effort	Relationships investigated^	Experimental design (highest rank of inference)*
Letnic et al. [122]	Relationships between dingoes and wildlife (arid)	Spatial replication of treatments Different measures for wildlife and dingoes Effect size measured	Replication devalued through seasonally staggered indexing Data influenced by seasonal and habitat differences in predator activity Used binary observations over potentially continuous measures Insensitive measure of grazing pressure used	25–30 plots over 25–30km (x2) 1 count at 8 sites over 2yrs	R3, R5	Quasi-experiment type I (5)
Lundie-Jenkins et al. [110]	Relationships between hare-wallabies and introduced mammals (arid)	Mensurative study Comprehensive dataset collected	Used binary observations over potentially continuous measures Non-independence between plots No details of dingo control program given Very small spatial scale	Intensive plot coverage within a ~10km² area 4 counts at 1 site over 1yr	R1, R2, R3, R4, R5, R6	Simple observations (16)
Moseby et al. [109]	Population dynamics of hopping-mice (arid)	Mensurative study Time-series data	Used binary observations over potentially continuous measures Very small spatial scale	4km transect inside an 8ha grid (x2) 15 counts at 2 sites over 8yrs	R3, R5	Quasi-experiment type II (6) or Pseudo-experiment type VI (14)
Newsome et al. [101]	Fence effect on dingoes and wildlife (arid)	Different measures for wildlife and dingoes	Invalid comparisons between species	Ringed plots around 10 waterpoints (x2) 4 counts at 1 site over 1yr	R3, R5	Quasi-experiment type 1 (5)
Pascoe [161]	Predator ecology and interactions (temperate)	Mensurative study Two measures of dingoes used Spatial replication	Used binary observations over potentially continuous measures for some analyses Sand plot index data untransformed	31 plots over 15km 8 counts at 3 sites over 2yrs	R2, R3, R5	Pseudo-experiment type V (13)
Pavey et al. [162]	Population dynamics of rodents and predators (arid)	Mensurative study Different measures for wildlife and dingoes Two measures of dingo abundance collected	Invalid assumptions when calculating the activity of predators Invalid comparisons between species Merged sandplot and spotlighting data	10km tracking transects (x3) 6 counts at 1 site over 2yrs	R3, R5	Pseudo-experiment type V (13)
Pettigrew [124]	The effect of dingo control on cats (arid)	Demographic data on cats collected Two measures of predators used	Ambiguous description of site and methodology Data from both sampling measures apparently combined Data from some treatments not reported	Spatial scale unknown, but ~100km of transect 12 counts at 1 site over 3yrs	R3, R4, R5	Quasi-experiment type IV (8)

Reference	Study topic (climate)	Methodological strengths	Methodological weaknesses	Spatial scale per site & sampling effort	Relation-ships investigated^	Experimental design (highest rank of inference)*
Purcell [123]	Dingo purity, diet, activity and behaviour (temperate)	Mensurative study Temporally intensive sampling	Used binary observations over potentially continuous measures for some analyses Sand plot index data untransformed	25 plots over 25km (x2) 26 counts at 1 site over 2yrs	R2, R3, R5	Pseudo-experiment type V (13)
Southgate et al. [103, 104]	Bilby and predator distribution and fire (arid)	Three different sampling strategies used Different measures of bilbies and predators	Data influenced by seasonal and habitat differences in predator activity Used binary observations over potentially continuous measures Invalid assumptions when calculating the activity of predators Footprints assumed 'old' were excluded from occupancy analysis	10km rectangle tracking transects (x2) 6–8 counts at 8 sites over 4yrs	R3, R5	Quasi-experiment type I (5)
Wallach & O'Neill [120] (a subset of [31, 78])	Relationship between dingoes and kowaris (arid)	Two measures of dingo abundance collected	Data influenced by seasonal and habitat differences in predator activity Invalid assumptions when calculating the relative abundance, "Index of abundance", and territorial activity of predators Data influenced by the presence of pet dogs and people Multiplication of binary and continuous abundance measures Sand plot index data untransformed Small spatial scale	10–12 strip plots (500m long), and 20 area plots (2ha) 1 count at 2 sites once	R2, R5	Quasi-experiment type IV (8)
Wallach et al. [163] (a subset of [31, 78])	Dingoes' role in protecting yellow-footed rock wallabies and malleefowl from predation by foxes and cats (arid, semi-arid)	Two measures of dingo abundance collected Large data set over wide spatial distribution	Data influenced by seasonal and habitat differences in predator activity Invalid assumptions when calculating the relative abundance, "Index of abundance", and territorial activity of predators Data influenced by the presence of pet dogs and people Multiplication of binary and continuous abundance measures Sand plot index data untransformed	9–25 strip plots (500m long), and 21–39 area plots (2ha) 1–2 counts at 7 sites over 1yr	R2, R5	Quasi-experiment type III (7)

Reference	Study topic (climate)	Methodological strengths	Methodological weaknesses	Spatial scale per site & sampling effort	Relation-ships investigated^	Experimental design (highest rank of inference)*
			Small spatial scale			
Wallach et al. [31]	The effect of dingo control on pack structure and social stability (arid)	Two measures of dingo abundance Large data set over wide spatial distribution	Data influenced by seasonal and habitat differences in predator activity Invalid assumptions when calculating the relative abundance, "Index of abundance", and territorial activity of predators Data influenced by the presence of pet dogs and people Multiplication of binary and continuous abundance measures Sand plot index data untransformed Small spatial scale	9–25 strip plots (500m long), and 21–39 area plots (2ha) 1–3 counts at 7 sites over 3yrs	R1	Quasi-experiment type III (7)
Wallach et al. [78]	The effect of dingo control on invasive species (arid)	Two measures of dingo abundance Large data set over wide spatial distribution	Data influenced by seasonal and habitat differences in predator activity Invalid assumptions when calculating the relative abundance, "Index of abundance", and territorial activity of predators Data influenced by the presence of pet dogs and people Multiplication of binary and continuous abundance measures Sand plot index data untransformed Small spatial scale	10–12 strip plots (500m long), and 20–40 area plots (2ha) 1–3 counts at 7 sites over 3yrs	R1, R4	Quasi-experiment type III (7)

Table 1. Methodological details of sand plot studies investigating the relationships between dingoes and faunal biodiversity. ^See Figure 1 for explanation of primary relationships. *See Table 1.2 in [89] for descriptions of experimental designs and rank of inference (rank 1 = highest possible, 16 = lowest possible). Note: different types of experimental design may be possible for some studies depending on the nature of the question/s being investigated, and the designs/rank identified here represent the highest level of design possible from the data collected.

4. The dingo-suppressive effects of foxes

The inability of correlations to describe causation was discussed by [68], and is illustrated here by examining published data on relationships between dingoes and foxes. Intraguild killing and interference competition are the two primary mechanisms given to facilitate the dominance of one predator over another ([1, 2], and references of studies therein). With some noteworthy exceptions (e.g. [144]), observations of intraguild killing are rare, and its occurrence is most often inferred from the remains of one predator in the diet of another (e.g. [164, 165]). Interference competition is typically inferred from studies of dietary overlap between sympatric predators (e.g. [118, 162, 166]), with high levels of dietary overlap used to infer a high level of potential competition. A variety of such studies have been conducted in Australia, which provide compelling correlative evidence that foxes may suppress dingoes through both mechanisms.

Dingo remains have been found in fox scats (e.g. [123, 164, 167, 168]), and even in cat scats (e.g. [169]), suggesting that these mesopredators kill (or at least consume) dingoes on some occasions. Being 2–3 times larger than foxes, dingoes will likely be victors in aggressive encounters between adults of the two species. However, foxes may be a threat to dingo pups, and dingoes may exhibit heightened activity levels during times when their pups are vulnerable [144]. By limiting recruitment of juveniles, foxes have been observed to suppress populations of one of Australia's largest native herbivores, eastern grey kangaroos M. giganteus [170]. Thus, differences in adult body sizes should not automatically discount the potential for foxes to suppress dingoes also. That mesopredators can slow down recruitment of top-predators was precisely the reason why smaller spotted hyaenas Crocuta crocuta were reintroduced with lions Panthera leo in southern Africa [171]. Multiple studies (e.g. [122, 164, 172, 173]) have also shown foxes to have a high level of dietary overlap with dingoes (Fig. 2), or in other words, dingoes and foxes eat the same things. This suggests that interference competition from high-density populations of foxes (which can reportedly be 7–20 times higher than dingoes [101]) reduces the availability of prey that otherwise might be consumed by dingoes; top-predators being primarily limited by bottom-up factors related to their preferred prey [174-176].

Figure 2. Ordination plot of nonmetric multidimensional scaling analyses showing a high level of dietary overlap between foxes (▼) and dingoes (▲) in the (A) Simpson Desert, (B) Strzelecki Desert and (C) Nullarbor region of arid Australia (from [164]).

Using data from [177], [178] report that dingoes were infrequently detected in places with high fox numbers (Fig. 3). This is further supported by the analyses of [136], which also report that dingo abundance is lower when fox abundance is high (Fig. 4). In contrast, scat indices (or scat collection rates) between dingoes and foxes appeared positively correlated in [123] and foxes (and especially goannas *Varanus varius*) were thought to derive some benefit from dingoes through kleptoparasitism in [173]. Although there are important limitations associated with the use of scats for making inferences about predation and abundance [16, 17, 61, 179], it appears clear from the data published in the aforementioned studies that a substantial and compelling amount of correlative evidence exists to support the hypothesis that foxes suppress dingoes through direct killing and interference competition. In all cases however, alternative hypotheses have been raised. These include the suppression of foxes by dingoes (e.g. [136, 164]) or the cumulative effect of livestock grazing (e.g. [15, 121]). That multiple plausible and competing alternative explanations can be generated is precisely the reason why correlative evidence cannot be trusted to describe causal processes [68] and most of the presently available literature on dingoes' ecological roles is at best inconclusive [52, 53]

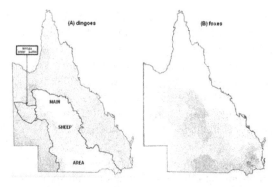

Figure 3. Bounty returns for (A) dingoes and (B) foxes in Queensland for the 1951–52 financial year (from [177], but see also [178]) showing that dingoes were rarely found in the presence of foxes.

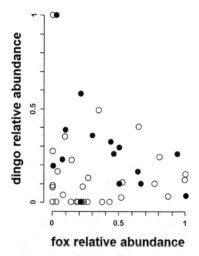

Figure 4. The relationship between dingo and fox abundance in eastern Australian forests (adapted from [136]) showing that the variability in dingo abundance is lower in areas with higher fox abundance (filled circles source data from [101], open circles source data from [157]).

Top-Predators as Biodiversity Regulators: Contemporary Issues Affecting Knowledge and Management of
Dingoes in Australia

77

5. What direct risk do dingoes pose to faunal biodiversity?

That dingoes provide net benefits to biodiversity has been almost universally accepted (e.g.
[9, 30, 47, 49, 62]) despite the unreliable and inconclusive state of the literature described
earlier. Additionally, and disregarded by most, is that dingoes have been implicated in the
extinctions of native vertebrates prior to European settlement [23, 180, 181] and the loss of
other native vertebrates in the recent past (e.g. [15, 19, 182-185]). Predation by dingoes and
other wild-living dogs is therefore identified as a known or potential threat in no less than
14 national threatened species recovery plans listed by the Australian government [17] for
species weighing as little as 70 g (i.e. marsupial moles, *Notoryctes* spp. [186]). 'Predation and
hybridisation by feral dogs (*Canis lupus familiaris*)' is also a listed Key Threatening Process
for 'threatened species, populations, and communities' in New South Wales (see [187] for
the listing, see [188] and [57] for the distribution of *Canis* sub-species in Australia, and see
[33, 189], [19], [56], [190], [22] for discussion of taxonomy and functional similarities between
wild-living sub-species of *Canis*). Dingoes also threaten northern hairy-nosed wombats
(*Lasiorhinus krefftii* [184, 191]), bridled nailtail wallabies (*Onychogalea fraenata* [146, 192]) and
a range of other species [16, 112, 193, 194] in other areas, where it is predicted that some
populations (such as those of koalas *Phascolarctos cinereus* [195, 196], for example) will only
persist through the control or absence of canid predators, including dingoes. Not only are
many mammals susceptible to exploitation by dingoes, but some bird (e.g. [19, 59, 197]) and
reptile (e.g. [112, 198-200]) populations may also be substantially impacted by them.
Predation on these less-preferred taxa may increase if mammals become increasingly
unavailable [16]. Urgent research focussing on R5 is therefore paramount before positive
dingo management is widely adopted in the hope that it will solve our biodiversity
conservation problems [16, 17].

Although dingoes and threatened native fauna coexisted sympatrically prior to European
settlement, they did not do so in the presence of rabbits, livestock or other landscape-
changing effects of pastoralism [23, 70, 201]. Unequivocal data on dingo densities may not
have been collected at the time, but post-European provision of virtually unlimited prey and
water resources across much of Australia has undoubtedly increased the range and
population densities of dingoes in areas outside the dingo barrier fence [19, 112, 202]. Thus,
populations of many native fauna have not been exposed to such high and ubiquitous
densities of dingoes until modern times. Put simply, the circumstances have changed
significantly since dingoes and now-threatened native fauna coexisted sustainably [15, 22],
where habitat alteration now enables dingoes (and other predators) to exploit populations
that otherwise might have sustained dingo predation. Thus, dingoes clearly present direct
risks to threatened fauna that must not be casually overlooked or assumed to be of lesser
importance than their indirect benefits [16, 17, 22]. For example, by applying established
predation risk assessment methods [50] developed for foxes and cats, [16] showed that up to
94% of extant threatened mammals, birds and reptiles in western New South Wales would
be at risk of dingo predation (71% at high risk) should dingoes re-establish there (Table 2).
By comparison, only 66% and 81% were predicted to be at risk of cat and fox predation [50].

	Low dingo density			High dingo density		
	No risk	Low risk	High risk	No risk	Low risk	High risk
EXTANT MAMMALS (n = 16)						
Vulnerable	4	2	2	0	2	6
Endangered	1	5	2	0	0	8
TOTAL	5	7	4	0	2	14
EXTANT BIRDS (n = 41)						
Vulnerable	16	13	2	4	12	15
Endangered	1	5	4	0	1	9
TOTAL	17	18	6	4	13	24
REPTILES (n = 23)						
Vulnerable	3	5	4	1	1	10
Endangered	2	5	4	0	2	9
TOTAL	5	10	8	1	3	19
LOCALLY EXTINCT MAMMALS (n = 17)						
TOTAL	2	6	9	2	0	15
LOCALLY EXTINCT BIRDS (n = 4)						
TOTAL	2	1	1	0	2	2

Table 2. Summary of overall dingo predation risks to 80 threatened extant and 21 locally extinct mammals, reptiles and birds in western New South Wales (from [16]).

Information on prey important to dingoes seems particularly useful for gauging the potential risks dingoes pose to threatened fauna [16]. While the mere presence of threatened species in dingo diets might be dismissed as uncommon events [169, 203, 204], 71% (33 of 47) of dingo diet studies assess <500 scat or stomach samples [17]. Greater sampling effort and a consideration of additional information has highlighted substantial risks to threatened fauna from dingoes in some cases (e.g. [17, 61, 112]). For example, threatened mammals under 35 g body weight are typically considered to fall outside the primary weight range [75, 205] of preferred prey for dingoes [19], but ([112]; N = 1907 scats) showed that anthropogenic provision of virtually unlimited food and water resources can exacerbate the risk of decline for some such species by facilitating elevated levels of dingo predation (i.e. hyperpredation [10, 206]). In another example, ([17]; N = 4087 scats) reported that although small rodents featured relatively infrequently in dingo scats while rabbits or kangaroos were available, consideration of dingo predation rates on rodents (made possible by knowledge of predator and prey densities) supported earlier assertions by [207] that dingoes alone have the capacity to exterminate rodent (e.g. dusky hopping-mice *Notomys fuscus*, Plate 1) populations within a few months under certain conditions, regardless of any

Top-Predators as Biodiversity Regulators: Contemporary Issues Affecting Knowledge and Management of
Dingoes in Australia

79

indirect benefit rodents may derive through dingoes' effects on foxes and cats [17]. Even seemingly unsusceptible arboreal and fossorial species (such as sugar gliders *Petaurus breviceps* and beach crabs *Ocypode* spp.) can become important prey for dingoes following the decline of their preferred prey ([61]; $N = 1460$ scats). Using the simple formula:

$$\text{Number of months until population extinction} = \frac{a}{\left(\dfrac{\left(b \times \left(365 \div 100\right)\right) \times c}{d} \right) \div 12}$$

where a = mean prey density, b = % occurrence of prey in scats, c = mean dingo pack size, and d = mean home range size of a dingo pack, the consideration of predator and prey densities can illuminate the significance of infrequent records of threatened species in dingo diets (Table 3).

Example	Dusky hopping-mice (from [17])	Rufous hare-wallabies (from [110])	Bridled nailtail wallabies (from [146])	Black-footed rock-wallabies (from [182])
Frequency of occurrence in dingo scats (%)	8*	12*	8*	46*
Mean dingo pack size (N=)	10*	10#	8^	5#
Mean dingo home range size (km²)	25*	50#	40^	50#
Prey density (individuals/km²)	60*	5#	5*	<1*
Predicted number of months until population extinction by dingoes	3.08	6.85	10.27	0.71

Table 3. The hypothetical impact of dingo predation on four threatened species based on the frequency of occurrence in dingo scats and predator and prey densities. (See [17] for rationale and assumptions; *Empirical data reported in original studies; ^L. Allen, unpublished data; #estimated values based on comparable studies).

As an example, [110] report the swift extinction of the small and last remaining mainland population (outside of fenced reserves) of rufous hare-wallabies *Lagorchestes hirsutus* (Plate 1) in 1987 when one or two foxes were detected first on only one occasion in an area that had just been exposed to a dingo control program. A cursory view of this outcome might suggest that dingo control facilitated the mesopredator release of foxes and led to the local extinction of a critically endangered species [41], but this does not explain the driver/s of hare-wallaby decline in the first place. Lethal dingo control had not previously occurred in the area until <100 poisoned baits were distributed along 20–30 km of vehicle tracks within the 10 km² area surrounding the hare-wallaby population (G. Lundie-Jenkins, unpublished data), so it could not have been lethal dingo control that caused the decline of the hare-wallabies. Foxes were reportedly absent (or at least uncommon [208]) until the dingo control program occurred [110], so it could not have been foxes which caused the decline either, and cats (which were also in very low abundance [110]) had probably been there for several decades [208, 209]. Notably, artificial water resources had not been established in the area until the 1950s and 1960s when outback mining and pastoralism became established [15, 112]. This undoubtedly increased the density and distribution of dingoes [112, 202] (the primary terrestrial predator of hare-wallabies since the extinction of thylacines [23]), suppressed any extant fox or cat populations, and caused or contributed to the decline of hare-wallabies and other marsupials [15, 19]. Furthermore, hare-wallabies were present in 12% of dingo scats collected prior to the commencement of the study [110]. Hare-wallaby densities were not reported in [110], but considering that the population became extinct just a few months later, there may have been only 50 or so animals (at most) in the population (G. Lundie-Jenkins, pers. comms.). If dingo densities were 0.2/km² (or 10 individuals within a home range of 50 km²) and hare-wallaby densities were 5/km² (or 50 individuals within the 10 km² study site), and assuming that one scat represents the prey eaten by a dingo in the previous 24 hours, then 12% occurrence in dingo scats could hypothetically represent as many as 438 hare-wallabies consumed by dingoes within the home range of a dingo pack each year. In other words, dingo predation alone had the capacity to exterminate the population of hare-wallabies in <7 months if they could not sustain the loss of that many individuals annually (Table 3). That dingoes were considered to be a limiting factor for their already endangered populations [110] (which is why lethal dingo control was initiated in the first place) suggests that, in association with other causal factors, increased dingo predation over the preceding 30–40 years (a consequence of adding water and dingo prey resources to the area) drove hare-wallabies down to a point where foxes just happened to be the predator to finish the extinction process.

In a somewhat comparable situation, [185] reported that one individual dingo in a dingo-controlled area (which was not detected on sand plots, but from post-mortem evidence on killed animals) was responsible for the surplus killing of 14 (out of 101) reintroduced (and similar sized) burrowing bettongs *Bettongia lesueur* on the first night after release, the rest succumbing to predation by unknown predators within a few months. It should also be noted that the simple calculations described earlier (in Table 3) falsely assume that predation rates remain constant as the prey population declines [17], which limit firm assertions from these considerations. But if the occurrence of a given species in dingo diets is known and a few key assumptions seem reasonable (discussed in [17]), then undertaking this coarse and hypothetical exercise can indicate whether or not dingoes should be considered a potential risk to the population before positive dingo management is implemented. From the preceding discussion, it should be clear that dingoes are certainly not the type of predator that one would want around a population of threatened fauna and should, as a precaution, be considered a significant threat until robust evidence suggests otherwise.

6. Practical issues hampering the realisation of net dingo benefits

Dingo suppression of mesopredators and herbivores are the two primary mechanisms predicted to generate positive biodiversity outcomes for fauna following positive dingo management (e.g. [23, 78]). Herbivore suppression is expected to increase the food and shelter available to threatened species, mesopredator suppression is expected to decrease predation on the same species, and dingoes are simply the tool expected to generate these outcomes. While the ecological theory supporting these mechanisms might be considered sound (e.g. [4, 6]; but see [210, 211] for an alternative considerations), at least two practical factors may prevent the realisation of these expected benefits in the rangelands of south-eastern Australia (where positive dingo management is considered imperative [50]).

6.1. Livestock enterprise switching

Sheep, goats, kangaroos and rabbits may be considered the most widespread and ecologically important herbivores in this area [34, 101, 212], but in places where two or more of them are extant, using dingoes to disentangle their cumulative impacts may be very difficult to achieve. Assuming that dingoes can suppress agriculturally non-productive herbivores (such as rabbits or kangaroos) without also suppressing the livestock with which they coexist, any reduction in undesirable herbivores may be replaced by increased stocking of agriculturally productive herbivores (such as sheep, goats or cattle), thereby maintaining total grazing pressure. For example, sheep populations have suffered precipitous declines in central and southern Queensland over the last decade [213], with no substantial change in the combined grazing pressure of sheep and cattle because of enterprise switching from sheep to cattle (Fig. 5), which are now in much higher densities in the area. Hence, enhancing the prospects for biodiversity conservation by securing improvements in vegetation communities might only be achievable if livestock stocking rates are not increased following the decline of some herbivores. But such may be a trivial consideration anyway, because dingoes are unlikely to kill only livestock competitors without also killing livestock [37, 189]. Importantly though, the positive management of dingoes may be advantageous to livestock producers where dingoes have greater effects on livestock competitors than they do on livestock ([39]; i.e. in arid cattle production regions), but this may not be economically or socially acceptable in places where the impacts of dingoes on smaller livestock species are prohibitive (i.e. sheep and goat production zones).

It should be understood that dingoes can completely eliminate sheep and goat populations [37, 44, 158, 212], and although their extirpation from rangelands might be considered a biodiversity success to some, the global human population need the food and fibre products these livestock produce [214-217]. As the world's largest wool exporter, the largest goat-meat exporter, and the second largest sheep-meat exporter (www.fao.org; www.mla.com.au), the loss of Australia as a globally important supplier of small ruminant products (which dingoes are quite capable of achieving [15, 61, 142, 218]) would need to be countered by an increase in livestock production in other countries. These countries may not be able to produce them as environmentally or economically sustainably as Australia; they may have extant diseases and other pathogens (such as rabies or screwworm flies *Cochliomyia* spp.) that inhibit broad-scale production or export, be forced to clear new land for increased livestock production, or may also have native predators of their own that need controlling in order to viably scale-up their production of livestock. In short, the primary reason for encouraging dingoes in sheep production areas (i.e. to improve biodiversity outcomes) may simply shift the biodiversity conservation problem to other countries where, unlike Australia, the extant top-predators may not be very common and their management may be more complex. These, and other issues will need serious consideration before dingoes are permitted to increase in sheep and goat production areas [22, 219].

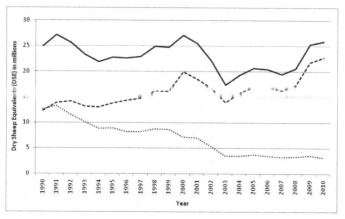

Figure 5. Trends in sheep (dotted line), cattle (dashed line; assuming 8 DSE per cow) and combined (solid line) livestock numbers in southwest and centralwest Queensland 1990–2010 (Australian Bureau of Statistics data, cat. no. 7121.0, Agricultural Commodities Australia, available at www.abs.gov.au).

6.2. Mesopredator release

Although many threatened fauna are indeed at risk of fox and cat predation [50], these fauna may also be equally at risk of dingo predation [16]. Dingoes do not kill only cats, foxes and kangaroos. In fact, these species are relatively uncommon in dingo diets [17, 19, 220], which means that replacing foxes and cats with dingoes (assuming dingoes could achieve this) or simply adding dingoes to an ecosystem might not stem the decline of threatened species [22]. As strongly interactive species, top-predators can have disproportionate effects on mesopredators, where small increases of larger predators dramatically reduce the abundance of smaller ones [1, 2]. Thus it is hypothetically conceivable that small increases in dingo abundances might substantially suppress foxes, leading to a net reduction in predator biomass and predation on threatened species. This does not appear to have been studied in great detail in Australia (Table 1) but may nevertheless prove true in some cases. Even so, the resulting lower levels of predation on threatened species might still be unsustainably high (which is why knowledge of R2 is of lesser value than R5 when considering the positive management of dingoes). In this situation, higher densities of dingoes might simply force threatened species to extinction slower than higher densities of mesopredators – the end result (extinction) being the same no matter which predator is most common (Table 3). Where multiple generalist predators are capable of exploiting the same prey species (as is the case with dingoes, foxes and cats [162, 164, 165, 172]), attempts to identify which predator is worse may be largely unhelpful in securing biodiversity against decline [221,

222]. Rather, identifying the population viability or status of threatened fauna under different management scenarios (R6) may be more useful.

A review of 14 cases of mesopredator release (analysed pairwise [223]) showed positive mesopredator population responses to decreases in higher-order predator abundance, suggesting that increases of dingoes might suppress foxes yet increase populations of cats, which are lower-order predators apparently suppressed by foxes [224]. Some support for this is found in several studies. Cats appeared to be positively associated with dingoes in the Tanami Desert of the Northern Territory [208], which is at the edge of foxes' national distribution [34, 99]. At tropical study sites devoid of foxes, [159] also reported that cats were positively associated with dingoes in the Northern Territory. At similar sites in the Kimberleys, [159] reported that (besides one outlier) cat activity varied little (0.18–0.40 tracks/sand plot/night) despite a nearly four-fold difference in dingo activity (0.80–4.30 tracks/sand plot/night). The cross-fence study of [121] (a subset of the data in [122]) also reported that foxes and cats were negatively and positively correlated with dingo presence, respectively, suggesting that increased dingoes may suppress foxes yet release cats from suppression by foxes. Subsequent analyses of the more comprehensive dataset suggested that cats were in equally low abundance on both sides of the fence [122], suggesting that cat abundance operated independently of the type of top-predator (dingoes or foxes) present.

Although increased populations of dingoes may reduce mesopredator activity they are unlikely to extirpate or exclude them (e.g. [118, 144, 225]). Detailed studies in northern South Australia ([225]; B. Allen, unpublished data from [32]) report the persistence of foxes in the presence of extremely high densities of dingoes, [144] reported that even though dingoes killed foxes they could not exclude them, and [118] showed that dingoes are unable to limit the distribution of foxes at landscape scales. Indeed, the colonisation and subsequent widespread distribution of foxes and cats across Australia [34] would suggest that the presence of dingoes (or the absence of lethal dingo control) neither prevented their establishment or limit their distribution. Rather, dingoes might reduce their densities and alter their behaviour at local scales [118], but whether or not this provides any relief to threatened prey remains unclear.

Given that dingoes are unlikely to extirpate cats, that there is strong overlap in the diets of dingoes, foxes and cats, and that cat predation is listed by the Australian Government as a Key Threatening Process to 18 of the 19 threatened arid-zone mammal species [122], there may be little overall biodiversity conservation benefit to species threatened by both foxes and cats if dingo populations increase [16, 22]. Irrespective of this, the positive management of dingoes would be unnecessary for places with extant (and typically unmanaged [32]) dingo populations, such as areas outside the dingo barrier fence, which are (confusingly) the very areas where some predict their positive management to be of most benefit to threatened fauna [122]. As illustrated earlier for rufous hare-wallabies and in addition to a variety of other important factors (discussed in [71, 72, 74, 226]), at-risk fauna are clearly threatened by predation per se, and not dingo *or* fox *or* cat predation individually (e.g. [221, 222, 227]). The literature is replete with examples of reductions of one pest animal increasing the undesirable impacts of another with no (or worse) overall outcomes for the species of conservation concern (e.g. [12, 110, 228]), and it would be naive to expect the positive management of dingoes across large areas to achieve universally 'good' outcomes for faunal biodiversity at more local scales [16, 22]. Increasing the number of generalist predators may only widen the suite of prey susceptible to predation and subsequent decline [222], and 'one may ask if the faunal biodiversity outcomes are any greater if a species is extinguished by a dingo instead of a fox or feral cat' [22]. Moreover, the biodiversity benefits expected of dingoes are likely to be available only to those prey species which have survived the impacts of cats, foxes and dingoes anyway. Thus, if fox and/or cat impacts are not the limiting factor for threatened species, then encouraging the suppression of foxes and cats by adding dingoes to the ecosystem seems an unlikely prerequisite for their recovery [16].

7. Context-specific management

Dingo impacts, roles and functions are context-specific, and the same is true for other top-predators [5, 229]. For example, the positive effects of wolves on biodiversity in some places may not be as apparent in other places just a few kilometres away, where site-specific factors may affect the strength of influence wolves have in the ecosystem [230, 231]. Such context-specific impacts mean that extreme caution should be exercised when considering using top-predators as biodiversity conservation tools in some new context, based on information collected from another time and place [22, 229]. Bottom-up factors associated with prey availability (such as habitat productivity, structural complexity etc) will affect the density of predators [174-176], the density of prey species [232-234] and their relative vulnerability to predation [221, 222, 227, 235]. Within this diversity, land use also varies from conservation to agriculture, from extensive to intensive livestock enterprises, and from small livestock to cattle production (e.g. [15, 69]). It is, therefore, unreasonable to expect that the goals and outcomes of dingo management will be uniform across Australia, which is why dingoes are presently managed locally for where they are and what they are (or are expected to be) doing [33, 35, 64].

Should positive dingo management to be adopted across large areas, the negative impacts of dingoes expected in some contexts may not be manageable in others. For example, the presence of dingoes has been predicted to benefit some rodents in arid environments [47], but dingo predation alone has the capacity to exterminate local populations of the same rodents under certain conditions (e.g. during droughts; Table 3; [17]) – conditions that are predicted to become more frequent and intense under future climate-change scenarios [236-238]. The negative impacts of dingoes in livestock production areas may also become increasingly unmanageable as dingoes are encouraged in adjacent conservation reserves where their impacts might be positive. Radio and GPS tracking studies indicate that most dingoes are sedentary (e.g. [108, 111, 239, 240]), and a recent continental-scale gene flow study [57] supports this conclusion. But a substantial proportion of dingoes do travel considerable distances (e.g. >550 km in 30 days [97]) for dispersal and exploration (e.g. [97, 123, 239, 241]). Given the capacity for dingoes to disperse, without containment fencing, dingo populations and their impacts (like reintroduced wolves [8]) are unlikely to remain only in reserves.

These issues are outside the capacity of any one individual or agency to manage, and are best addressed through a strategic adaptive management approach that can accommodate differences in situation and objectives [242-244]. The management of dingoes (either positively or negatively) requires adherence to a number of underlying principles including: defining the biological assets to be protected and the people involved, setting measurable goals and timeframes for action, undertaking management actions at a scale appropriate to the enterprise or ecosystem to be enhanced and the wild dog home range and movements, relying on a suite of actions applied in a coordinated sequence, and continuously monitoring in preparation for new incursions or threats [35, 64]. Issues of scale and management unit are particularly important, and the minimum size of the management unit may be determined using the home range size of the animal in the particular environment as a guide. Recorded home range sizes for dingoes vary from 7–2013 km^2 in semi-arid and arid rangeland rangelands, from 2–262 km^2 in mesic environments, and may be <1 km^2 in urban areas [19, 112, 239, 245]. Such variation in scales important to dingoes is likely to preclude management approaches which seek to apply broad-scale solutions to context-dependant problems, such as the widespread prohibition of dingo control for the recovery of an isolated population of threatened mammals.

Although dingo management policies must be general by nature, the process of defining the issue in strategic management ensures that the appropriate scale for actions is decided before commencement. Therefore, where dingoes are determined by reliable experimentation to be important for biodiversity conservation, strategic management can achieve this objective locally or regionally, depending on the minimum size of the management unit required. In short, top-down management approaches which seek to exclude the land manager in favour of government policy intervention (e.g. [70]) and/or apply broad-scale solutions to context-dependant impacts (either positive or negative) are unlikely to succeed in restoring faunal biodiversity [22, 246].

8. Looking forward: surmountable challenges to overcome

Knowing that the available data is lacking rigour and defensible or definite conclusions may seem depressing after the countless hours of hard work expended by many in obtaining it. But all is not lost, and dismissing it completely may be just as dangerous as embracing it uncritically [53]. From the implications of [52], [95] and the present study it seems clear that a greater understanding of the advantages and limitations of sand plot tracking indices are required by many dingo researchers, and it will be difficult in reaching consensus on the state of the available literature until this is achieved. The advantages and limitations of indices and populations estimation procedures have been widely discussed (in [67, 93, 94, 105, 106, 114-116, 247-249]; to cite just a few) to a point where relative abundance indices can be viewed as an incredibly powerful population censusing technique provided appropriate principles and analyses are applied [93, 114]. Moreover, so long as the results of studies with lower inferential ability are valued above those with designs that permit more definitive statements, end-users of the literature may also continue to be confused about the most appropriate dingo and threatened species management strategies. A return to more objective and applied science and management of dingoes is imperative (also suggested by [189]).

Long-term manipulative experiments are able to advance science much more rapidly than other approaches [68, 89, 90], but they are few (Table 1), and more are sorely needed [41, 250]. When conducting such studies, the relationships (Fig. 1) and knowledge gaps being investigated are of utmost importance. Interest in the positive management of dingoes as biodiversity conservation tools is ultimately driven by the desire to improve the status of threatened fauna through trophic effects (e.g. [23, 50]), so should not the threatened faunal response to dingo management be the variable of interest? Demonstration of sustained non-target population responses to predator control can provide 'conclusive proof' [79] for the effects of lethal dingo control on threatened fauna. Hence, in places where dingoes are actively controlled (for whatever reason), it is not the direct or indirect effects of dingoes on fauna that should be of primarily interest, but rather, the effects of dingo management practices on fauna (R6) – the 'black box' approach [86]. Knowledge of the other relationships (R2, R3, R5) is supplementary and may be more important in places where dingoes are typically unmanaged.

In order to focus our collective attention on the questions that matter most, we issue the following challenge. For any given site and population of threatened species:

1. Do contemporary dingo management practices negatively affect the species either directly or indirectly?
2. Do dingoes themselves pose a current or future threat to the species, regardless of their indirect effects on other threatening processes?
3. Is positive dingo management the only practical option to improve conditions for the species?
4. What factors determine which predator becomes ecologically dominant following dingo control programs?

If contemporary dingo management practices (such as poison baiting, trapping or shooting) do not harm threatened species either directly or indirectly (R6), then arguments to cease controlling dingoes remain unjustified on biodiversity conservation grounds. Multiple studies have failed to demonstrate the 'release' of mesopredators following dingo control (R4) (e.g. [87, 88, 159, 251], and no studies to date have shown short-term negative responses from populations of non-target species to dingo or fox control [79]. Hence, lethal dingo control will still be useful in mitigating livestock losses without fear of releasing mesopredators or harming threatened species. If dingoes threaten a particular species to any degree (R5), then researchers must investigate the relative strengths of dingo-prey (R5), mesopredator-prey (R3), and dingo-mesopredator (R2) interactions in order to gauge the likely outcomes of positive dingo management. Positive dingo management is unlikely to benefit the threatened species where the direct effect of dingoes is greater (or may become greater) than their indirect effect on mesopredators.

If dingo control does appear to hinder the conservation of the species, and dingoes do not pose a current or future threat to them, are there any alternative management actions that could improve biodiversity outcomes without compromising livestock production values? For example, livestock guardian dogs might offer a non-lethal approach to reduce the impacts of dingoes on livestock without excluding dingoes from an area [252, 253].

Top-Predators as Biodiversity Regulators: Contemporary Issues Affecting Knowledge and Management of
Dingoes in Australia

87

Alternatively, the selective exclusion of agriculturally non-productive herbivores from watering points [254-256] may elicit a greater bottom-up response from threatened species than the top-down suppression of mesopredators by dingoes without threatening the viability of livestock producers. In fact, doing so would probably enhance their viability.

Lastly, the commonly observed presence of foxes in areas free of dingo control suggests that bottom-up factors may largely determine which predator successfully colonises and dominates an area, though these influences remain largely unknown. Foxes appear to be positively associated with disturbed agricultural habitats in a bottom-up manner [257, 258], which may help explain the pattern of fox densities noted by [178] and others (e.g. [50]). Top-predators can also be associated with higher biodiversity in a bottom-up manner [19, 174, 175, 229], and positive correlations between dingoes and greater biodiversity values cannot be immediately interpreted to be the result of top-down processes [52, 68]. When the factors that determine which predator dominates a given area become well understood, our ability to manage predators will be greatly enhanced.

9. Conclusion

Maintaining top-predator function may be an important component of biodiversity conservation initiatives in many places [1, 2]. Although this might be more easily achieved in relatively intact areas, the functions of top-predators may be most needed in the more degraded ecosystems characterised by depleted faunal and floral communities. Importantly though, such systems are typically those used most heavily by humans for agricultural production, and the age-old battle betweens humans and top-predators seems likely to continue into the foreseeable future [214, 259]. Nevertheless, conservative environmental management is required in our efforts to balance the needs of humans with those of the threatened fauna and flora we seek to protect [260]. Evidence-based biodiversity conservation and carefully considered policy approaches are critical to the informed management of top-predators for this purpose [261, 262].

This chapter has discussed the knowledge and management of dingoes for biodiversity conservation. Our overview of the field data underpinning knowledge of dingoes' ecological roles has identified critical knowledge gaps that we believe require the primary attention of researchers and policy makers operating in this area. We have also shown that although dingoes are well-studied, their functional roles may not be well understood. This is because methodological flaws, sampling bias and experimental design limitations inherent to most studies (Table 1; [52]) cannot provide reliable or conclusive evidence for dingoes ecological roles. We therefore agree with [53] that there is inconclusive evidence for the positive roles of dingoes and that cessation of lethal dingo control is presently unjustified on biodiversity conservation grounds. We are cognizant that questioning the conclusions of studies documenting the benefits of fox control on native fauna [263] probably delayed the necessary implementation of broad-scale fox control for biodiversity conservation in many places. Likewise, we acknowledge that questioning the science underpinning the role of dingoes may delay the adoption of positive dingo management in places that might yet be shown to need it. However, we believe there are sufficient concerns regarding the impacts of dingoes on mesopredators and threatened fauna to stress strong caution when considering the positive management of dingoes for biodiversity conservation purposes under current ecological conditions [22].

We therefore challenge researchers and funding agencies to focus on applied science questions that can address the effects of dingo management practices on prey populations of interest. Doing so within an experimental framework that has the capacity to explore and exclude alternative hypotheses will be most useful, and we encourage those with such data to invest time in its analyses and publication. We encourage the continued interest in dingoes as a biodiversity conservation tool, and look forward to the results of future studies on this charismatic and iconic terrestrial top predator.

Author details

Benjamin L. Allen* and Luke K-P. Leung
School of Agriculture and Food Sciences, the University of Queensland, Gatton, Queensland, Australia

Peter J.S. Fleming
Vertebrate Pest Research Unit, Department of Primary Industries, Orange, New South Wales , Australia

Matt Hayward
Centre for African Conservation Ecology, Nelson Mandela Metropolitan University, Port Elizabeth, South Africa
School of Biological, Earth and Environmental Science, University of New South Wales, Sydney, Australia

Lee R. Allen
Robert Wicks Pest Animal Research Centre, Biosecurity Queensland, Toowoomba, Queensland, Australia

Richard M. Engeman
National Wildlife Research Centre, US Department of Agriculture, Fort Collins, Colorado, USA

Guy Ballard
Vertebrate Pest Research Unit, Department of Primary Industries, Armidale, New South Wales, Australia

Acknowledgement

We thank the editors for the invitation to prepare this report, which was enhanced by discussions, correspondence and input from a variety of researchers, land managers and policy makers involved in dingo and threatened species research and management.

* Corresponding Author

10. References

[1] Ray JC, Redford KH, Steneck RS, Berger J. Large carnivores and the conservation of biodiversity. Washington: Island Press; 2005.

[2] Terborgh J, Estes JA. Trophic cascades: Predator, prey, and the changing dynamics of nature. Washington D.C.: Island Press; 2010.

[3] Miller BJ, Harlow HJ, Harlow TS, Biggins D, Ripple WJ. Trophic cascades linking wolves (*Canis lupus*), coyotes (*Canis latrans*), and small mammals. Canadian Journal of Zoology. 2012;90:70-8.

[4] Ripple WJ, Beschta RL. Trophic cascades in Yellowstone: The first 15 years after wolf reintroduction. Biological Conservation. 2012;145(1):205-13.

[5] Soule ME, Estes JA, Miller B, Honnold DL. Strongly interacting species: conservation policy, management, and ethics. BioScience. 2005 2005/02/01;55(2):168-76.

[6] Estes JA, Terborgh J, Brashares JS, Power ME, Berger J, Bond WJ, et al. Trophic downgrading of planet earth. Science. 2011;333:301-6.

[7] Hayward MW, Somers MJ. Reintroduction of top-order predators. Oxford: Wiley-Blackwell; 2009.

[8] Bangs EE, Smith DW. Re-introduction of the gray wolf into Yellowstone National Park and central Idaho, USA. In: Soorae PS, editor. Global re-introduction perspectives: re-introduction case studies from around the globe. Abu Dhabi, UAE: IUCN/SSC Re-introduction Specialist Group; 2008.

[9] Ritchie EG, Elmhagen B, Glen AS, Letnic M, Ludwig G, McDonald RA. Ecosystem restoration with teeth: what role for predators? Trends in Ecology and Evolution. 2012;27(5):265-71.

[10] Courchamp F, Langlais M, Sugihara G. Rabbits killing birds: Modelling the hyperpredation process. Journal of Animal Ecology. 2000;69(1):154-64.

[11] Nogales M, Martin A, Tershy B, Donlan CJ, Veitch CR, Puerta N, et al. A review of feral cat eradication on islands. Conservation Biology. 2004;18:310-9.

[12] Ruscoe WA, Ramsey DSL, Pech RP, Sweetapple PJ, Yockney I, Barron MC, et al. Unexpected consequences of control: competitive vs. predator release in a four-species assemblage of invasive mammals. Ecology Letters. 2011;14(10):1035-42.

[13] Engeman R, Whisson D, Quinn J, Cano F, Quiñones P, White Jr. TH. Monitoring invasive mammalian predator populations sharing habitat with the critically endangered Puerto Rican parrot *Amazona vittata*. Oryx. 2006;40(1):95-102.

[14] Miller B, Reading R, Forest S. Prairie night: Black-footed ferrets and the recovery of endangered species. Washington DC: Smithsonian Institution Press; 1996.

[15] Allen BL. A comment on the distribution of historical and contemporary livestock grazing across Australia: Implications for using dingoes for biodiversity conservation Ecological Management and Restoration. 2011;12(1):26-30.

[16] Allen BL, Fleming PJS. Reintroducing the dingo: The risk of dingo predation to threatened species in western New South Wales. Wildlife Research. 2012;39(1):35-50.

[17] Allen BL, Leung LK-P. Assessing predation risk to threatened fauna from their prevalence in predator scats: dingoes and rodents in arid Australia. PLoS ONE. 2012;7(5):e36426.

[18] Gompper ME. Top carnivores in the suburbs? Ecological and conservation issues raised by colonization of northeastern North America by coyotes. BioScience. 2002;52(2):185-90.

[19] Corbett LK. The dingo in Australia and Asia. Second ed. Marleston: J.B. Books, South Australia; 2001.

[20] Carbone C, Mace GM, Roberts SC, Macdonald DW. Energetic constraints on the diet of terrestrial carnivores. Nature. [10.1038/46266]. 1999;402(6759):286-8.

[21] Carbone C, Teacher A, Rowcliffe JM. The costs of carnivory. PloS Biology. 2007;5(2: e22):0363-8.

[22] Fleming PJS, Allen BL, Ballard G. Seven considerations about dingoes as biodiversity engineers: the socioecological niches of dogs in Australia. Australian Mammalogy. 2012;01(1):119-01.

[23] Johnson C. Australia's mammal extinctions: A 50 000 year history. Melbourne: Cambridge University press; 2006.

[24] Johnson CN, Wroe S. Causes of extinction of vertebrates during the Holocene of mainland Australia: Arrival of the dingo, or human impact? Holocene. 2003 Nov;13(6):941-8.

[25] Letnic M, Fillios M, Crowther MS. Could direct killing by larger dingoes have caused the extinction of the thylacine from mainland Australia? PLoS ONE. 2012;7(5):e34877.

[26] Oskarsson MCR, Klutsch CFC, Boonyaprakob U, Wilton A, Tanabe Y, Savolainen P. Mitochondrial DNA data indicate an introduction through Mainland Southeast Asia for Australian dingoes and Polynesian domestic dogs. Proceedings of the Royal Society B. In press;xx(xx):xx-xx.

[27] Saetre P, Lindberg J, Leonard JA, Olsson K, Pettersson U, Ellegren H, et al. From wild wolf to domestic dog: Gene expression changes in the brain. Molecular Brain Research. 2004 2004/7/26;126(2):198-206.

[28] Savolainen P, Leitner T, Wilton AN, Matisoo-Smith E, Lundeberg J. A detailed picture of the origin of the Australian dingo, obtained from the study of mitochondrial DNA. Proceedings of the National Academy of Sciences of the United States of America. 2004 August 17, 2004;101(33):12387-90.

[29] vonHoldt BM, Pollinger JP, Lohmueller KE, Han E, Parker HG, Quignon P, et al. Genome-wide SNP and haplotype analyses reveal a rich history underlying dog domestication. Nature. 2010;464:898-903.

[30] Bowman D. Bring elephants to Australia? Nature. 2012;482(7383):30.

[31] Wallach AD, Ritchie EG, Read J, O'Neill AJ. More than mere numbers: The impact of lethal control on the stability of a top-order predator. PloS ONE. 2009;4(9):e6861.

[32] Allen BL. The effect of lethal control on the conservation values of Canis lupus dingo. In: Maia AP, Crussi HF, editors. Wolves: Biology, conservation, and management. New York: Nova Publishers; 2012. p. 79-108.

[33] Allen BL, Engeman RM, Allen LR. Wild dogma II: The role and implications of wild dogma for wild dog management in Australia. Current Zoology. 2011;57(6):737-40.

[34] West P. Assessing invasive animals in Australia 2008. Canberra: National Land and Water Resources Audit, The Invasive Animals Cooperative Research Centre2008.

[35] Fleming PJS, Allen BL, Allen LR, Ballard G, Bengsen AJ, Gentle MN, et al. Management of wild canids in Australia: free-ranging dogs and red foxes. In: Glen AS, Dickman CR, editors. Carnivores of Australia: past, present and future. Collingwood: CSIRO Publishing; In press.

[36] Hewitt L. Major economic costs associated with wild dogs in the Queensland grazing industry. Brisbane: Agforce2009.

[37] Thomson PC. Dingoes and sheep in pastoral areas. Journal of Agriculture. 1984;25:27-31.

[38] Allen BL, editor. The effect of regional dingo control on calf production in northern South Australia, 1972-2008. Queensland Pest Animal Symposium; 2010; Gladstone, Queensland.

[39] Wicks S, Allen BL, editors. Returns on investment in wild dog management: beef production in the South Australian arid lands. 56th Australian Agricultural Resource Economics Society conference; 2012; Fremantle, Western Australia: AARES.

[40] Glen AS, Dickman CR. Complex interactions among mammalian carnivores in Australia, and their implications for wildlife management. Biological Reviews. 2005 August;80(3):387-401.

[41] Glen AS, Dickman CR, Soule ME, Mackey BG. Evaluating the role of the dingo as a trophic regulator in Australian ecosystems. Austral Ecology. 2007;32(5):492-501.

[42] Barnes TS, Goldizen AW, Morton JM, Coleman GT. Cystic echinococcosis in a wild population of the brush-tailed rock-wallaby (Petrogale penicillata), a threatened macropodid. Parasitology. 2008;135:715-23.

[43] Jenkins DJ, Macpherson CNL. Transmission ecology of Echinococcus in wild-life in Australia and Africa. Parasitology. 2003 2003;127(Supplement):S63-S72.

[44] Allen LR, Sparkes EC. The effect of dingo control on sheep and beef cattle in Queensland. Journal of Applied Ecology. 2001;38:76-87.

[45] Atkinson SA, editor. Dingo control or conservation? Attitudes towards urban dingoes (Canis lupus dingo) as an aid to dingo management. 23rd Vertebrate Pest Conference; 2008; California: University of California, Davis.

[46] Hytten KF. Dingo dualisms: Exploring the ambiguous identity of Australian dingoes. Australian Zoology. 2009;35(1):18-27.

[47] Letnic M, Ritchie EG, Dickman CR. Top predators as biodiversity regulators: the dingo Canis lupus dingo as a case study. Biological Reviews. 2012;87(2):390-419.

[48] Visser RL, Watson JEM, Dickman CR, Southgate R, Jenkins D, Johnson CN. Developing a national framework for dingo trophic regulation research in Australia: Outcomes of a national workshop. Ecological Management and Restoration. 2009;10(2):168-70.

[49] Carwardine J, O'Connor T, Legge S, Mackey B, Possingham HP, Martin TG. Priority threat management to protect Kimberley wildlife. Brisbane: CSIRO Ecosystem Sciences; 2011.

[50] Dickman C, Glen A, Letnic M. Reintroducing the dingo: Can Australia's conservation wastelands be restored? In: Hayward MW, Somers MJ, editors. Reintroduction of top-order predators. Oxford: Wiley-Blackwell; 2009. p. 238-69.

[51] Allen BL. Did dingo control cause the elimination of kowaris through mesopredator release effects? A response to Wallach and O'Neill (2009). Animal Biodiversity and Conservation. 2010;32(2):1-4.

[52] Allen BL, Engeman RM, Allen LR. Wild dogma I: An examination of recent "evidence" for dingo regulation of invasive mesopredator release in Australia. Current Zoology. 2011;57(5):568-83.

[53] Glen AS. Enough dogma: seeking the middle ground on the role of dingoes. Current Zoology. In press;xx(xx):xx-xx.

[54] Letnic M, Crowther MS, Dickman CR, Ritchie E. Demonising the dingo: How much wild dogma is enough? Current Zoology. 2011;57(5):668-70.

[55] Jones E. Hybridisation between the dingo, Canis lupus dingo, and the domestic dog, Canis lupus familiaris, in Victoria. A critical review. Australian Mammology. 2009;31:1-7.

[56] Corbett LK. Canis lupus ssp. dingo. IUCN 2010. IUCN Red List of Threatened Species. Version 2010.4. www.iucnredlist.org. 2008:Downloaded on 20 April 2011.

[57] Stephens D. The molecular ecology of Australian wild dogs: hybridisation, gene flow and genetic structure at multiple geographic scales [PhD]. Perth: The University of Western Australia; 2011.

[58] Elledge AE, Allen LR, Carlsson B-L, Wilton AN, Leung LK-P. An evaluation of genetic analyses, skull morphology and visual appearance for assessing dingo purity: implications for dingo conservation. Wildlife Research. 2008;35:812-20.

[59] Boland CRJ. Breeding biology of Rainbow Bee-Eaters (Merops ornatus): A migratory, colonial, cooperative bird. Auk. 2004 Jul;121(3):811-23.

[60] Smith BP, Litchfield CA. A review of the relationship between indigenous Australians, dingoes (Canis dingo) and domestic dogs (Canis familiaris). Anthrozoös. 2009;22(2):111-28.

[61] Allen LR, Goullet M, Palmer R. The diet of the dingo (Canis lupus dingo and hybrids) in north-eastern Australia: a supplement to Brook and Kutt. The Rangeland Journal. In press;xx(xx):xx-xx.

[62] Baker L, Letnic M, Nesbitt B. Ecological functional landscapes: the role of native top-order predators as trophic regulators. New South Wales: Public submission to the Northern Rivers Catchment Management Authority; 2011.

[63] Clarke M. Final recommendation on a nomination for listing: Canis lupus subsp. dingo. Victoria: Scientific Advisory Committee, Department of Sustainability and Environment; 2007.

[64] Fleming P, Corbett L, Harden R, Thomson P. Managing the impacts of dingoes and other wild dogs. Bomford M, editor. Canberra: Bureau of Rural Sciences; 2001.

[65] Ritchie E, editor. Dingoes: managing for ecosystem resilience and pastoral productivity. Queensland Pest Animal Symposium; 2010; Gladstone.

[66] Linnell JDC. The relative importance of predators and people in structuring and conserving ecosystems. Conservation Biology. 2011;25(3):646-7.

[67] Caughley G. Analysis of vertebrate populations. London: John Wiley and Sons; 1977.

[68] MacKenzie DI, Nichols JD, Royle JA, Pollock KH, Bailey LL, Hines JE. Occupancy estimation and modelling: Inferring patterns and dynamics of species occurrence. London: Academic Press (Elsevier); 2006.

[69] Hamblin A. Land, Australia State of the environment report (Theme report). Canberra: CSIRO Publishing, on behalf of the Department of Environment and Heritage; 2001.

[70] Letnic M. Dispossession, degradation, and extinction: environmental history in arid Australia. Biodiversity and Conservation. 2000;9:295-308.

[71] McKenzie NL, Burbidge AA, Baynes A, Brereton RN, Dickman CR, Gordon G, et al. Analysis of factors implicated in the recent decline of Australia's mammal fauna. Journal of Biogeography. 2007;34(4):597-611.

[72] Dickman CR, Pressey RL, Lim L, Parnaby HE. Mammals of particular conservation concern in the Western Division of New South Wales. Biological Conservation. 1993 1993;65:219-48.

[73] Lunney D. Causes of the extinction of native mammals of the western division of New South Wales: An ecological interpretation of the nineteenth century historical record. Rangeland Journal. 2001;23(1):44-70.

[74] Smith PJ, Pressey RL, Smith JE. Birds of particular conservation concern in the western division of New South Wales. Biological Conservation. 1994;69:315-38.

[75] Burbidge AA, McKenzie NL. Patterns in the modern decline of Western Australia's vertebrate fauna: Causes and conservation implications. Biological Conservation. 1989;50:143-98.

[76] Purcell BV. Dingo. *Australian Natural History Series*. Collingwood: CSIRO Publishing; 2010.

[77] Fleming PJS, Allen LR, Lapidge SJ, Robley A, Saunders GR, Thomson PC. Strategic approach to mitigating the impacts of wild canids: Proposed activities of the Invasive Animals Cooperative Research Centre. Australian Journal of Experimental Agriculture. 2006;46(6-7):753-62.

[78] Wallach AD, Johnson CN, Ritchie EG, O'Neill AJ. Predator control promotes invasive dominated ecological states. Ecology Letters. 2010;13:1008-18.

[79] Glen AS, Gentle MN, Dickman CR. Non-target impacts of poison baiting for predator control in Australia. Mammal Review. 2007;37(3):191-205.

[80] Elledge AE, Leung LK, Allen LR, Firestone K, Wilton AN. Assessing the taxonomic status of dingoes *Canis familiaris dingo* for conservation. Mammal Review. 2006;36(2):142-56.

[81] Denny EA, Dickman CR. Review of cat ecology and management strategies in Australia. Canberra: Invasive Animals Cooperative Research Centre; 2010.

[82] Saunders GR, Gentle MN, Dickman CR. The impacts and management of foxes *Vulpes vulpes* in Australia. Mammal Review. 2010;40(3):181-211.

[83] Allen LR. Best-practice baiting: Evaluation of large-scale, community-based 1080 baiting campaigns. Toowoomba: Robert Wicks Pest Animal Research Centre, Department of Primary Industries (Biosecurity Queensland); 2006.

[84] APVMA. Review findings for sodium monofluoroacetate: The reconsideration of registrations of products containing sodium monofluoroacetate and approvals of their associated labels, Environmental Assessment. Canberra: Australian Pesticides and Veterinary Medicines Authority; 2008.

[85] Visser RL, Watson JEM, Dickman CR, Southgate R, Jenkins D, Johnson CN. A national framework for research on trophic regulation by the Dingo in Australia. Pacific Conservation Biology. 2009;15:209-16.

[86] Kinnear JE, Krebs CJ, Pentland C, Orell P, Holme C, Karvinen R. Predator-baiting experiments for the conservation of rock-wallabies in Western Australia: A 25-year review with recent advances. Wildlife Research. 2010;37:57-67.

[87] Allen LR. The impact of wild dog predation and wild dog control on beef cattle production, PhD Thesis. Brisbane: Department of Zoology, The University of Queensland; 2005.

[88] Eldridge SR, Shakeshaft BJ, Nano TJ. The impact of wild dog control on cattle, native and introduced herbivores and introduced predators in central Australia, Final report to the Bureau of Rural Sciences. Alice Springs: Parks and Wildlife Commission of the Northern Territory2002.

[89] Hone J. Wildlife damage control. Collingwood, Victoria: CSIRO Publishing; 2007.

[90] Platt JR. Strong inference: Certain systematic methods of scientific thinking may produce much more rapid progress than others. Science. 1964;146(3642):347-53.

[91] Meek PD. Refining and improving the use of camera trap technology for wildlife management and research in Australia and New Zealand Final report to the Winston Churchill Memorial Trust of Australia. 2011.

[92] Robley A, Woodford L, Lindeman M, Ivone G, Beach M, Campbell I, et al. Assessing the effectiveness of buried baiting for the control of wild dogs in Victoria. Heidelberg, Victoria: Department of Sustainability and Environment; 2011.

[93] Engeman R. Indexing principles and a widely applicable paradigm for indexing animal populations. Wildlife Research. 2005;32(3):202-10.

[94] Wilson GJ, Delahay RJ. A review of methods to estimate the abundance of terrestrial carnivores using field signs and observation. Wildlife Research. 2001;28:151-64.

[95] Mahon PS, Banks PB, Dickman CR. Population indices for wild carnivores: A critical study in sand-dune habitat, south-western Queensland. Wildlife Research. 1998;25:11-22.

[96] Creel S, Christianson D. Relationships between direct predation and risk effects. Trends in Ecology and Evolution. 2008;23(4):194-201.

[97] Allen LR. Best practice baiting: dispersal and seasonal movement of wild dogs (*Canis lupus familiaris*). Technical highlights: Invasive plant and animal research 2008-09. Brisbane: QLD Department of Employment, Economic Development and Innovation; 2009. p. 61-2.

[98] Cogger H. Reptiles and amphibians of Australia (Sixth edition). Florida: Ralph Curtis Publishing; 2000.

[99] Saunders G, McLeod L. Improving fox management strategies in Australia. Canberra: Bureau of Rural Sciences; 2007.

[100] Evangelista P, Engeman R, Tallents L. Testing a passive tracking index for monitoring the endangered Ethiopian wolf. Integrative Zoology. 2009;4:172-8.

[101] Newsome AE, Catling PC, Cooke BD, Smyth R. Two ecological universes separated by the dingo barrier fence in semi-arid Australia: Interactions between landscapes,

Top-Predators as Biodiversity Regulators: Contemporary Issues Affecting Knowledge and Management of
Dingoes in Australia

95

herbivory and carnivory, with and without dingoes. Rangeland Journal. 2001;23(1):71-98.

[102] Edwards GP, de Preu N, Crealy IV, Shakeshaft BJ. Habitat selection by feral cats and dingoes in a semi-arid woodland environment in central Australia. Austral Ecology. 2002;27:26-31.

[103] Southgate R, Paltridge R, Masters P, Carthew S. Bilby distribution and fire: a test of alternative models of habitat suitability in the Tanami Desert, Australia. Ecography. 2007 December;30(6):759-76.

[104] Southgate R, Paltridge R, Masters P, Ostendorf B. Modelling introduced predator and herbivore distribution in the Tanami Desert, Australia. Journal of Arid Environments. 2007 Feb;68(3):438-64.

[105] Sutherland WJ. Ecological census techniques. Cambridge, UK: Cambridge University Press; 1996.

[106] Pollock KH. The challenge of measuring change in wildlife populations: A biometrician's perspective. In: Grigg GC, Hale PT, Lunney D, editors. Conservation through the sustainable use of wildlife. The University of Queensland: Centre for Conservation Biology; 1995. p. 117-21.

[107] Krebs CJ. Ecology: The experimental analysis of distribution and abundance. 6 ed. San Francisco: Benjamin-Cummings Publishing; 2008.

[108] Thomson PC. The behavioural ecology of dingoes in north-western Australia: IV. Social and spatial organisation, and movements. Wildlife Research. 1992;19(5):543-63.

[109] Moseby KE, Owens H, Brandle R, Bice JK, Gates J. Variation in population dynamics and movement patterns between two geographically isolated populations of the dusky hopping mouse (Notomys fuscus). Wildlife Research. 2006;33:223-32.

[110] Lundie-Jenkins G, Corbett LK, Phillips CM. Ecology of the rufous hare-wallaby, Lagorchestes hirsutus Gould (Marsupialia: Macropodidae), in the Tanami Desert, Northern Territory. III. Interactions with introduced mammal species. Wildlife Research. 1993;20:495-511.

[111] Allen BL. Do desert dingoes drink daily? Visitation rates at remote waterpoints in the Strzelecki Desert. Australian Mammalogy. In press;xx(xx):xx-xx.

[112] Newsome TM. Ecology of the dingo (Canis lupus dingo) in the Tanami Desert in relation to human-resource subsidies [PhD]. Sydney: The University of Sydney; 2011.

[113] Buckland ST, Anderson DR, Burnham KP, Laake JL, Borchers DL, Thomas LN. Introduction to distance sampling: estimating abundance of biological populations. Oxford: Oxford University Press; 2001.

[114] Johnson DH. In defense of indices: the case of bird surveys. Journal of Wildlife Management. 2008;72(4):857-68.

[115] Anderson DR. The need to get the basics right in wildlife field studies. Wildlife Society Bulletin. 2001;29(4):1294-7.

[116] Engeman RM. More on the need to get the basics right: population indices. Wildlife Society Bulletin. 2003;31:286-7.

[117] Atwood TC, Fry TL, Leland BR. Partitioning of anthropogenic watering sites by desert carnivores. The Journal of Wildlife Management. 2011;75(7):1609-15.

[118] Mitchell BD, Banks PB. Do wild dogs exclude foxes? Evidence for competition from dietary and spatial overlaps. Austral Ecology. 2005;30(5):581-91.

[119] Hayward MW, Slotow R. Temporal partitioning of activity in large African carnivores: tests of multiple hypotheses. South African Journal of Wildlife Research. 2009;39:109-25.

[120] Wallach AD, O'Neill AJ. Threatened species indicate hot-spots of top-down regulation. Animal Biodiversity and Conservation. 2009;32(2):127-33.

[121] Letnic M, Crowther M, Koch F. Does a top-predator provide an endangered rodent with refuge from a mesopredator? Animal Conservation. 2009;12(4):302-12.

[122] Letnic M, Koch F, Gordon C, Crowther M, Dickman C. Keystone effects of an alien top-predator stem extinctions of native mammals. Proceedings of the Royal Society of London B. 2009;276:3249-56.

[123] Purcell DW. Order in the pack. Ecology of Canis lupus dingo in the southern Greater Blue Mountains World Heritage Area [PhD]. Sydney: University of Western Sydney, School of Natural Sciences; 2009.

[124] Pettigrew JD. A burst of feral cats in the Diamantina: a lesson for the management of pest species? In: Siepen G, Owens C, editors. Cat Management Workshop Proceedings. Brisbane: Queensland Department of Environment and Heritage; 1993. p. 25-32.

[125] Brown JS, Laundre JW, Gurung M. The ecology of fear: optimal foraging, game theory, and trophic interactions. Journal of Mammalogy. 1999; 80:385-99.

[126] Ripple WJ, Beschta RL. Wolves and the ecology of fear: Can predation risk structure ecosystems? Bioscience. 2004;54(8):755-66.

[127] Engeman RM, Pipas MJ, Gruver KS, Allen LR. Monitoring coyote population changes with a passive activity index. Wildlife Research. 2000;27:553-7.

[128] Karanth KU, Chundawat RS, Nichol JD, Kumar NS. Estimation of tiger densities in the tropical dry forests of Panna, Central India, using photographic capture-recapture sampling. Animal Conservation. 2004 Aug;7:285-90.

[129] Karanth KU, Nichols JD. Estimation of tiger densities in India using photographic captures and recaptures. Ecology Letters. 1998;79:2852-62.

[130] Karanth KU. Estimating tiger *Panthera tigris* populations from camera-trap data using capture-recapture models. Biological Conservation. 1995;71(3):333-8.

[131] Karanth KU, Kumar NS, Nichols JD. Field surveys: Estimating absolute densities of tigers using capture-recapture sampling. Monitoring tigers and their prey: A manual for researchers, managers and conservationists in tropical Asia. Karnataka: Center for Wildlife Studies; 2002.

[132] Rowcliffe JM, Field J, Turvey ST, Carbone C. Estimating animal density using camera traps without the need for individual recognition. Journal of Applied Ecology. 2008;45:1228-36.

[133] Dawson MJ, Miller C. Aerial mark–recapture estimates of wild horses using natural markings. Wildlife Research. 2008;35(4):365-70.

[134] Grigg GC, Beard LA, Alexander P, Pople AR, Cairns SC. Aerial survey of kangaroos in South Australia 1978-1998: A brief report focusing on methodology. Australian Zoologist. 1999 June;31(1):292-300.

[135] Stephens PA, Zaumyslova OJ, Miquelle D, Myslenkov AE, Hayward GD. Estimating population density from indirect sign: track counts and the Formozov-Malyshev-Pereleshin formula. Animal Conservation. 2006;9:339-48.

[136] Johnson C, VanDerWal J. Evidence that dingoes limit the abundance of a mesopredator in eastern Australian forests. Journal of Applied Ecology. 2009;46:641-6.

[137] Johnson CN, Isaac JL, Fisher DO. Rarity of a top predator triggers continent-wide collapse of mammal prey: Dingoes and marsupials in Australia. Proceedings of the Royal Society, Biological Sciences Series B. 2007 February 7;274(1608):341-6.

[138] Letnic M, Koch F. Are dingoes a trophic regulator in arid Australia? A comparison of mammal communities on either side of the dingo fence. Austral Ecology. 2010;35(2):267–175.

[139] Fleming PJS, Thompson JA, Nicol HI. Indices for measuring the efficacy of aerial baiting for wild dog control in north-eastern New South Wales. Wildlife Research. 1996;23(6):665-74.

[140] Brawata RL, Neeman T. Is water the key? Dingo management, intraguild interactions and predator distribution around water points in arid Australia. Wildlife Research. 2011;38(5):426-36.

[141] Chippendale JF. The Queensland barrier fence - A preliminary economic analysis. Proceedings of the Vertebrate Pest Control Conference. Adelaide1991. p. 143-7.

[142] Payne WLR, Fletcher JW, Tomkins B. Report on the Royal Commission on rabbit, dingo, and stock route administration. Queensland: Queensland Government; 1930.

[143] Yelland L. Holding the line: A history of the South Australian Dog Fence Board, 1947 to 2000. Adelaide: Primary Industries and Resources South Australia; 2001.

[144] Moseby KE, Neilly H, Read JL, Crisp HA. Interactions between a top order predator and exotic mesopredators in the Australian rangelands. International Journal of Ecology. 2012;Article ID 250352:15 pages.

[145] White GC. Why take calculus? Rigour in wildlife management. Wildlife Society Bulletin. 2001;29:380-6.

[146] Augusteyn J, editor. Determining the effectiveness of canine control at Taunton National park (Scientific) and its impact on the population of bridled nailtail wallabies. Queensland Pest Animal Symposium; 2010; Gladstone, Queensland.

[147] Burrows ND, Algar D, Robinson AD, Sinagra J, Ward B, Liddelow G. Controlling introduced predators in the Gibson Desert of Western Australia. Journal of Arid Environments. 2003;55:691-713.

[148] Catling PC, Burt RJ. Studies of the ground-dwelling mammals of eucalypt forests in south-eastern New South Wales: The effect of habitat variables on distribution and abundance. Wildlife Research. 1995;22:271-88.

[149] Catling PC, Hertog A, Burt RJ, Wombey JC, Forrester RI. The short-term effect of cane toads (Bufo marinus) on native fauna in the Gulf Country of the Northern Territory. Wildlife Research. 1999;26(2):161-85.

[150] Christensen P, Burrows N. Project desert dreaming: experimental reintroduction of mammals to the Gibson Desert, Western Australia. In: Serena M, editor. Reintroduction biology of Australian and New Zealand fauna. Sydney: Surrey Beatty & Sons; 1995. p. 199-207.

[151] Claridge AW, Cunningham RB, Catling PC, Reid AM. Trends in the activity levels of forest-dwelling vertebrate fauna against a background of intensive baiting for foxes. Forest Ecology and Management. 2010;260(5):822-32.

[152] Corbett L. Does dingo predation or buffalo competition regulate feral pig populations in the Australian wet-dry tropics? An experimental study. Wildlife Research. 1995;22(1):65-74.

[153] Edwards GP, Dobbie W, Berman DM. Warren ripping: Its impacts on European rabbits and other wildlife of central Australia amid the establishment of rabbit haemorrhagic disease. Wildlife Research. 2002;29(6):567-75.

[154] Edwards GP, Dobbie W, Berman DM. Population trends of European rabbits and other wildlife of central Australia in the wake of rabbit haemorrhagic disease. Wildlife Research. 2002,29.557-63.

[155] Fillios M, Gordon C, Koch F, Letnic M. The effect of a top predator on kangaroo abundance in arid Australia and its implications for archaeological faunal assemblages. Journal of Archaeological Science. 2010;37(5):986-93.

[156] Fleming PJS. Aspects of the management of wild dogs (Canis familiaris) in north-eastern New South Wales. Armidale: The University of New England; 1996.

[157] Catling PC, Burt RJ. Why are red foxes absent from some eucalypt forests in eastern New South Wales? Wildlife Research. 1995;22:535-46.

[158] Newsome A. The biology and ecology of the dingo. In: Dickman C, Lunney D, editors. A symposium on the dingo: Royal Zoological Society of New South Wales; 2001.

[159] Kennedy M, Phillips B, Legge S, Murphy S, Faulkner R. Do dingoes suppress the activity of feral cats in northern Australia? Austral Ecology. 2011;37(1):134-9.

[160] Koertner G, Watson P. The immediate impact of 1080 aerial baiting to control wild dogs on a spotted-tailed quoll population. Wildlife Research. 2005;32(8):673-80.

[161] Pascoe JH. Apex predators in the Greater Blue Mountains World Heritage Area [PhD]. Sydney: The University of Western Sydney; 2011.

[162] Pavey CR, Eldridge SR, Heywood M. Population dynamics and prey selection of native and introduced predators during a rodent outbreak in arid Australia. Journal of Mammalogy. 2008 June;89(3):674-83.

[163] Wallach AD, Murray BR, O'Neill AJ. Can threatened species survive where the top predator is absent? Biological Conservation. 2009;142:43-52.

[164] Cupples JB, Crowther MS, Story G, Letnic M. Dietary overlap and prey selectivity among sympatric carnivores: could dingoes suppress foxes through competition for prey? Journal of Mammalogy. 2011;92(3):590-600.

[165] Glen AS, Pennay M, Dickman CR, Wintle BA, Firestone KB. Diets of sympatric native and introduced carnivores in the Barrington Tops, eastern Australia. Austral Ecology. 2011;36(3):290-6.

[166] Letnic M, Dworjanyn SA. Does a top predator reduce the predatory impact of an invasive mesopredator on an endangered rodent? Ecography. 2011;34(5):827-35.

[167] Triggs B, Brunner H, Cullen JM. The food of fox, dog and cat in Croajingalong National Park, south-eastern Victoria. Australian Wildlife Research. 1984;11:491-9.

Top-Predators as Biodiversity Regulators: Contemporary Issues Affecting Knowledge and Management of Dingoes in Australia

99

[168] Lunney D, Law B, Rummery C. Contrast between visible abundance of the brush-tailed rock wallaby, *Petrogale penicillata*, and its rarity in fox and dog scats in the gorges east of Armidale, New South Wales. Wildlife Research. 1996;23(3):373-80.

[169] Paltridge R. The diets of cats, foxes and dingoes in relation to prey availability in the Tanami Desert, Northern Territory. Wildlife Research. 2002 2002;29:389-403.

[170] Banks PB, Newsome AE, Dickman CR. Predation by red foxes limits recruitment in populations of eastern grey kangaroos. Austral Ecology. 2000;25:283-91.

[171] Hayward MW, Adendorff J, O'Brien J, Sholto-Douglas A, Bissett C, Moolman LC, et al. The reintroduction of large carnivores to the Eastern Cape Province, South Africa: an assessment. Oryx. 2007;41:205-14.

[172] Glen AS, Dickman CR. Niche overlap between marsupial and eutherian carnivores: Does competition threaten the endangered spotted-tailed quoll? Journal of Applied Ecology. 2008 April;45(2):700-7.

[173] Pascoe JH, Mulley RC, Spencer R, Chapple R. Diet analysis of mammals, raptors and reptiles in a complex predator assemblage in the Blue Mountains, eastern Australia. Australian Journal of Zoology. 2012;59(5):295-301.

[174] Carbone C, Gittleman JL. A common rule for the scaling of carnivore density. Science. 2002;295:2273-6.

[175] Hayward MW, O'Brien J, Kerley GIH. Carrying capacity of large African predators: predictions and tests. Biological Conservation. 2007;139:219-29.

[176] Jedrzejewska B, Jedrzejewski W. Predation in vertebrate communities: the Bialowieza Primeval Forest as a case study. Berlin, Germany: Springer; 1998.

[177] Anon. Annual report on the operations of 'The Stock Routes and Rural Lands Protection Acts, 1944-1951' for the year 1951-52. Queensland Parliamentary Papers. 1952;1951-52.

[178] Letnic M, Greenville A, Denny E, Dickman CR, Tischler M, Gordon C, et al. Does a top predator suppress the abundance of an invasive mesopredator at a continental scale? Global Ecology and Biogeography. 2011;20(2):343-53.

[179] Mitchell B, Balogh S. Monitoring techniques for vertebrate pests: Wild dogs. Orange: NSW Department of Primary Industries, Bureau of Rural Sciences; 2007.

[180] Archer M. New information about the Quaternary distribution of the thylacine (Marsupialia, Thylacinidae) in Australia. Journal of the Proceedings of the Royal Society of Western Australia. 1974; 57:43-50.

[181] Baird RF. The dingo as a possible factor in the disappearance of the *Gallinula mortierii* from the Australian mainland. Emu. 1991;91:121-2.

[182] Moseby K, Read J, Gee P, Gee I. A study of the Davenport Range black-footed rock wallaby colony and possible threatening processes. Final report to Wildlife Conservation Fund, Department for Environment and Heritage. 1998:Adelaide.

[183] Kerle JA, Foulkes JN, Kimber RG, Papenfus D. The decline of the brushtail possum, *Trichosurus vulpecula* (Kerr 1798), in arid Australia. The Rangeland Journal. 1992;14(2):107-27.

[184] Horsup A. Recovery plan for the northern hairy-nosed wombat *Lasiorhinus krefftii* 2004-2008 Canberra: Report produced by the Environmental Protection

Agency/Queensland Parks and Wildlife Service for the Department of Environment and Heritage; 2004.

[185] Moseby KE, Read JL, Paton DC, Copley P, Hill BM, Crisp HA. Predation determines the outcome of 10 reintroduction attempts in arid South Australia. Biological Conservation. 2011;144(12):2863-72.

[186] Benshemesh J. Recovery plan for marsupial moles *Notoryctes typhlops* and *N. caurinus*, 2005-2010. Alice Springs: Northern Territory Department of Infrastructure, Planning and Environment; 2004.

[187] Major R. Predation and hybridisation by feral dogs (*Canis lupus familiaris*) - Key threatening process listing: New South Wales Department of Environment, Climate Change, and Water; 2009.

[188] Corbett LK. The conservation status of the dingo *Canis lupus dingo* in Australia, with particular reference to New South Wales: threats to pure dingoes and potential solutions. In: Dickman CR, Lunney D, editors. Symposium on the dingo. Mossman: Royal Zoological Society of New South Wales, Australian Museum; 2001. p. 10-9.

[189] Coman B, Jones E. The loaded dog: On objectivity in the biological sciences and the curious case of the dingo. Quadrant. 2007;1 Nov 2007.

[190] Claridge A, Hunt R. Evaluating the role of the Dingo as a trophic regulator: Additional practical suggestions. Ecological Management and Restoration. 2008;9(2):116-9.

[191] Banks SC, Horsup A, Wilton AN, Taylor AC. Genetic marker investigation of the source and impact of predation on a highly endangered species. Molecular Ecology. 2003 June 2003;12(6):1663-7.

[192] Lundie-Jenkins G, Lowry J. Recovery plan for the bridled nailtail wallaby (*Onychogalea fraenata*) 2005-2009: Report to the Department of Environment and Heritage (DEH), Canberra. Brisbane: Environmental Protection Agency/Queensland Parks and Wildlife Service2005.

[193] Coutts-Smith AJ, Mahon PS, Letnic M, Downey PO. The threat posed by pest animals to biodiversity in New South Wales. Canberra: Invasive Animals Cooperative Research Centre2007.

[194] Newsome A, Pech R, Smyth R, Banks P, Dickman C. Potential impacts on Australian native fauna of rabbit calicivirus disease. Canberra: Biodiversity Group, Environment Australia1997.

[195] Lunney D, Gresser S, O'Neill LE, Matthews A, Rhodes J. The impact of fire and dogs on koalas at Port Stephens, New South Wales, using popualtion viability analysis. Pacific Conservation Biology. 2007;13(3):189-201.

[196] Mifsud G. Hansard excerpt of the transcript from the Commonwealth senate enquiry into the status, health and sustainability of Australia'a koala population, 19 May 2011. Canberra: Official Committee Hansard, the Commonwealth of Australia; 2011.

[197] Benshemesh J. National recovery plan for malleefowl. South Australia: Department for Environment and Heritage; 2007.

[198] Heard GW, Robertson P, Black D, Barrow G, Johnson P, Hurley V, et al. Canid predation: a potentially significant threat to relic populations of the Inland Carpet Python *Morelia spilota metcalfei* (Pythonidae) in Victoria. The Victorian Naturalist. 2006;123(2):68-74.

Top-Predators as Biodiversity Regulators: Contemporary Issues Affecting Knowledge and Management of Dingoes in Australia

101

[199] Somaweera R, Webb JK, Shine R. It's a dog-eat-croc world: dingo predation on the nests of freshwater crocodiles in tropical Australia. Ecological Research. 2011;26(5):957-67.

[200] Whiting SD, Long JL, Hadden KM, Lauder ADK, Koch AU. Insights into size, seasonality and biology of a nesting population of the Olive Ridley turtle in northern Australia. Wildlife Research. 2007;34(3):200-10.

[201] James CD, Landsberg J, Morton SR. Provision of watering points in the Australian arid zone: A review of effects on biota. Journal or Arid Environments. 1999;41:87-121.

[202] Davies KF, Melbourne BA, James CD, Cuningham RB. Using traits of species to understand responses to land use change: Birds and livestock grazing in the Australian arid zone. Biological Conservation. 2010;143:78-85.

[203] Brook LA, Kutt AS. The diet of the dingo (Canis lupus dingo) in north-eastern Australia with comments on its conservation implications. The Rangeland Journal. 2011;33:79-85.

[204] Claridge AW, Mills DJ, Barry SC. Prevalence of threatened native species in canid scats from coastal and near-coastal landscapes in south-eastern Australia. Australian Mammalogy. 2010;32(2):117-26.

[205] Johnson CN, Isaac JL. Body mass and extinction risk in Australian marsupials: The 'Critical Weight Range' revisited. Austral Ecology. 2009;34(1):35-40.

[206] Smith AP, Quin DG. Patterns and causes of extinction and decline in Australian conilurine rodents. Biological Conservation. 1996;77:243-67.

[207] Newsome AE, Corbett LK. Outbreaks of rodents in semi-arid and arid Australia: Causes, preventions, and evolutionary considerations. In: Prakash I, Gosh PK, editors. Rodents in desert environments. The Hague, The Netherlands: Dr. W. Junk; 1975.

[208] Paltridge RM. Predator-prey interactions in the spinifex grasslands of central Australia [PhD]. Wollongong: School of Biological Sciences, University of Wollongong; 2005.

[209] Abbott I. Origin and spread of the cat, Felis catus, on mainland Australia, with a discussion of the magnitude of its early impact on native fauna. Wildlife Research. 2002;29:51-74.

[210] Wilson DS. The adequacy of body size as a niche difference. American Naturalist. 1975;109:769-84.

[211] Hutchinson GE. Homage to Santa Rosalina, or why are there so many kinds of animals? American Naturalist. 1959;93:145-59.

[212] Pople A, Froese J. Distribution, abundance and harvesting of feral goats in the Australian rangelands 1984-2011. Final report to ACRIS Management Committee. Brisbane: Department of Employment, Economic Development and Innovation; 2012.

[213] East IJ, Foreman I. The structure, dynamics and movement patterns of the Australian sheep industry. Australian Veterinary Journal. 2011;89(12):477-89.

[214] Gordon IJ, Acevedo-Whitehouse K, Altwegg R, Garner TWJ, Gompper ME, Katzner TE, et al. What the 'food security' agenda means for animal conservation in terrestrial ecosystems. Animal Conservation. 2012;15(2):115-6.

[215] Kearney J. Food consumption trends and drivers. Philosophical Transactions of The Royal Society B. 2010;365:2793-807.

[216] Thornton PK. Livestock production: recent trends, future prospects. Philosophical Transactions of The Royal Society B. 2010;365:2853-67.

[217] Wright HL, Lake IR, Dolman PM. Agriculture—a key element for conservation in the developing world. Conservation Letters. 2012;5(1):11-9.

[218] Allen LR. Control of dingoes inside the Barrier Fence. Rural Lands Protection Board meeting minutes. 1987;May:40-2.

[219] Rural Management Partners. Economic assessment of the impact of dingoes/wild dogs in Queensland. Queensland: Commissioned by the Department of Natural Resources2004. Report No.: Project LP02/03NRM.

[220] Corbett L. Does dingo predation or buffalo competition regulate feral pig populations in the Australian wet-dry tropics? An experimental study. Wildlife Research. 1995 1995;22:65-74.

[221] Holmes JC. Population regulation: A dynamic complex of interactions. Wildlife Research. 1995;22:11-9.

[222] Holt RD, Lawton JH. The ecological consequences of shared natural enemies. Annual Review of Ecology and Systematics. 1994;25: 495-520.

[223] Ritchie EG, Johnson CN. Predator interactions, mesopredator release and biodiversity conservation. Ecology Letters. 2009;12(9):982-98.

[224] Risbey DA, Calver MC, Short J, Bradley JS, Wright IW. The impact of cats and foxes on the small vertebrate fauna of Heirisson Prong, Western Australia. II. A field experiment. Wildlife Research. 2000;27(3):223-35.

[225] Bird P. Improved electric fences and baiting techniques: A behavioural approach to integrated dingo control. Adelaide: Animal and Plant Control Commission, Department of Primary Industries South Australia; 1994.

[226] Woinarski JCZ, Legge S, Fitzsimons JA, Traill BJ, Burbidge AA, Fisher A, et al. The disappearing mammal fauna of northern Australia: context, cause, and response. Conservation Letters. 2011;4(3):192-201.

[227] Sinclair ARE, Pech RP, Dickman CR, Hik D, Mahon P, Newsome AE. Predicting the effects of predation on conservation of endangered prey. Conservation Biology. 1998;12(3):564-75.

[228] de Tores PJ, Marlow NJ. A review of the relative merits of predator exclusion fencing and repeated 1080 fox baiting for protection of native fauna: five case studies from Western Australia. In: Somers MJ, Hayward MW, editors. Fencing for conservation. New York: Springer; 2011. p. 21-42.

[229] Sergio F, Caro T, Brown D, Clucas B, Hunter J, Ketchum J, et al. Top predators as conservation tools: Ecological rationale, assumptions, and efficacy. Annual Review of Ecology, Evolution and Systematics. 2008;39:1-19.

[230] Garrott RA, Gude JA, Bergman EJ, Gower C, White PJ, Hamlin KL. Generalizing wolf effects across the Greater Yellowstone Area: a cautionary note. Wildlife Society Bulletin. 2005 2005/12/01;33(4):1245-55.

[231] Vucetich JA, Hebblewhite M, Smith DW, Peterson RO. Predicting prey population dynamics from kill rate, predation rate and predator–prey ratios in three wolf-ungulate systems. Journal of Animal Ecology. 2011;80(6):1236-45.

[232] Dickman CR, Letnic M, Mahon PS. Population dynamics of two species of dragon lizards in arid Australia: the effects of rainfall. Oecologia. 1999;119:357-66.

Top-Predators as Biodiversity Regulators: Contemporary Issues Affecting Knowledge and Management of Dingoes in Australia

103

[233] Dickman CR, Mahon PS, Masters P, Gibson DF. Long-term dynamics of rodent populations in arid Australia: The influence of rainfall. Wildlife Research. 1999;26:389-403.

[234] Pople AR, Grigg GC, Phinn SR, Menke N, McAlpine C, Possingham HP. Reassessing the spatial and temporal dynamics of kangaroo populations. In: Coulson G, Eldridge M, editors. Macropods: The biology of kangaroos, wallabies and rat-kangaroos. Melbourne: CSIRO Publishing; 2010. p. 197-210.

[235] Gese EM, Knowlton FF. The role of predation in wildlife population dynamics. In: Ginnetr TF, Henke SE, editors. The role of predator control as a tool in game management. San Angelo, Texas: Texas Agricultural Research and Extension Center; 2001. p. 7-25.

[236] Whetton PH, Fowler AM, Haylock MR, Pittock AB. Implications of climate change due to the enhanced greenhouse effect on floods and droughts in Australia. Climatic Change. 1993;25(3):289-317.

[237] Moise AF, Hudson DA. Probabilistic predictions of climate change for Australia and southern Africa using the reliability ensemble average of IPCC CMIP3 model simulations. Journal of Geophysical Research. 2008;113:D15113.

[238] Hughes L. Climate change and Australia: trends, projections and impacts. Austral Ecology. 2003;28:423-43.

[239] Claridge AW, Mills DJ, Hunt R, Jenkins DJ, Bean J. Satellite tracking of wild dogs in south-eastern mainland Australian forests: Implications for management of a problematic top-order carnivore. Forest Ecology and Management. 2009;258:814-22.

[240] Robertshaw JD, Harden RH. The ecology of the dingo in north-eastern New South Wales, I. Movement and home range. Australian Wildlife Research. 1985;12(1):25-37.

[241] Thomson PC, Rose K, Kok NE. The behavioural ecology of dingoes in north-western Australia: VI. Temporary extraterritorial movements and dispersal. Wildlife Research. 1992;19(5):585-95.

[242] Braysher M. Managing vertebrate pests: Principles and strategies. Canberra: Bureau of Rural Sciences, Australian Government Publishing; 1993.

[243] Chapple RS, Ramp D, Bradstock RA, Kingsford RT, Merson JA, Auld TD, et al. Integrating science into management of ecosystems in the Greater Blue Mountains. Environmental Management. 2011;48:659-74.

[244] Holling CS. Adaptive environmental assessment and management. London: John Wiley and Sons; 1978.

[245] Allen BL, editor. The spatial ecology and zoonoses of urban dingoes, and the use of Traversed Area Polygons (TAPs) to calculate home range sizes. workshop on the remote monitoring of wild canids and felids; 2007 21-22 March 2007; Australian National University, Canberra: Invasive Animals Cooperative Research Centre.

[246] Sodhi NS, Butler R, Raven PH. Bottom-up conservation. Biotropica. 2011;43(5):521-3.

[247] Caughley G, Sinclair ARE. Wildlife ecology and management. Cambridge, Massachusetts: Blackwell Sciences; 1994.

[248] Engeman R, Pipas M, Gruver K, Bourassa J, Allen L. Plot placement when using a passive tracking index to simultaneously monitor multiple species of animals. Wildlife Research. 2002;29(1):85-90.

[249] Engeman RM, Allen LR, Zerbe GO. Variance estimate for the activity index of Allen *et al.* Wildlife Research. 1998;25(6):643-8.

[250] Lindenmayer DB, Likens GE, Andersen A, Bowman D, Bull M, Burns E, et al. Value of long-term ecological studies. Austral Ecology. In press;xx(xx):xx-xx.

[251] Fleming PJS. Ground-placed baits for the control of wild dogs: Evaluation of a replacement-baiting strategy in north-eastern New South Wales. Wildlife Research. 1996;23(6):729-40.

[252] van Bommel L, Johnson CN. Good dog! Using livestock guardian dogs to protect livestock from predators in Australia's extensive grazing systems. Wildlife Research. 2012;39(3):220-9.

[253] Allen LR, Byrne D. How do guardian dogs 'work'? 15th Australasian Vertebrate Pest Conference; Sydney, Australia: Invasive Animal Cooperative Research Centre; 2011. p 158.

[254] Norbury GL. An electrified watering trough that selectively excludes kangaroos. The Rangeland Journal. 1992;14(1):3-8.

[255] Berman D, Brennan M, Elsworth P. How can warren destruction by ripping control European wild rabbits (*Oryctolagus cuniculus*) on large properties in the Australian arid zone? Wildlife Research. 2011;38(1):77-88.

[256] Russell BG, Letnic M, Fleming PJS. Managing feral goat impacts by manipulating their access to water in the rangelands. The Rangeland Journal. 2011;33(2):143-52.

[257] Killengreen ST, Lecomte N, Ehrich D, Schott T, Yoccoz NG, Ims RA. The importance of marine vs. human-induced subsidies in the maintenance of an expanding mesocarnivore in the arctic tundra. Journal of Animal Ecology. 2011;80(5):1049-60.

[258] Elmhagen B, Rushton SP. Trophic control of mesopredators in terrestrial ecosystems: top-down or bottom-up? Ecology Letters. 2007;10:197-206.

[259] Treves A, Karanth KU. Human-carnivore conflict and perspectives on carnivore management worldwide. Conservation Biology. 2003;17(6):1491-9.

[260] Cooney R. The precautionary principle in biodiversity conservation and natural resource management: An issues paper for policy-makers, researchers, and practicioners. Gland, Switzerland and Cambridge: IUCN; 2004.

[261] Trouwborst A. Managing the carnivore comeback: international and EU species protection law and the return of lynx, wolf and bear to western Europe. Journal of Environmental Law. 2010;22(3):347-72.

[262] Pullin AS, Knight TM. Doing more good than harm: building an evidence-base for conservation and environmental management. Biological Conservation. 2009;142:931-4.

[263] Hone J. Fox control and rock-wallaby population dynamics - assumptions and hypotheses. Wildlife Research. 1999;26:671-3.

Chapter 4

Species-Diversity Utilization of Salt Lick Sites at Borgu Sector of Kainji Lake National Park, Nigeria

A. G. Lameed and Jenyo-Oni Adetola

Additional information is available at the end of the chapter

1. Introduction

Mineral elements occur in the living tissues or soil in either large or small quantities. Those that occur in large quantities are called macro/major elements while those that occur in small quantities are called micro/minor/trace elements. These macro elements are required in large amount and the micro are required in small amount (Underwood, 1977, Alloway, 1990,) They occur in the tissues of plants and animals in varied concentrations. The magnitude of this concentration varies greatly among different living organisms and part of the organisms (W.B.E, 1995).

Although most of the naturally occurring mineral elements are found in the animal tissues, many are present merely because they are constituents of the animal's food and may not have essential function in the metabolism of the animals. Hence essential mineral elements refer to as mineral elements are those which had been proven to have a metabolic role in the body (McDonald, 1987). The essential minerals elements are necessary to life for work such as enzyme and hormone metabolisms (W.B.E, 1995). Enzymes are activated by trace elements known as metallo enzymes (Mertz, 1996). Ingestion or uptake of minerals that are deficient, inbalanced, or excessively high in a particular mineral element induces changes in the functions, activities, or concentration of that element in the body tissue or fluids. Biochemical defects develop, physiological functions are affected and structural disorder may arise and death may occur under the circumstances (Pethes, 1980).

The influence of wildlife management with the goal of maintaining wildlife population and the entire biodiversity at maximum level and maximum ecosystem utilization depend heavily on the knowledge of mineral elements in the nutrients requirement of animals (Mertz, 1976).

2. Salt lick/mineral lick

Salt licks are deposit of mineral salts used by animals to supplement their nutrition, ensuring enough minerals in their diets. A wide assortment of animals, primarily herbivores use salt licks to get essential nutrients like calcium magnesium, sodium and zinc. Salt licks are natural mineral which are mineral outcrops in the soil which are visited by herbivores for soil eating (biting and chewing) or licking (with tongue). They also supplement mineral that are deficient in animal vegetable diets (Ayeni, 1972).

Animals regularly visit licks in the ecosystem which are composed of primarily common salt (sodium chloride). It provides sodium, calcium, iron, phosphorus etc (Ayeni, 1972). Salt licks occur naturally in certain locations in the forest where mineral salt are found on the ground surface.

Shortage of sodium in the plants which are eaten by wildlife could motivate the game to eat a lot of soil at the lick (Ayeni, 1972). The shortage is as a result of water soluble sodium salts being leached out during heavy rain following long period of desiccation. Some plants even substitute potassium ions for sodium ions uptake from soil without showing mineral deficiency symptom (Buckman and Brandy 1960). Many plants are also rich in sodium and potassium, example are; *Acacia elatoir* (Brenan), *Achyranthes aspera (L)*, *Calyptotheca premna resinosa (Hochst)* and *Salvadora persica (L)* and *Echolium amplexicaule* (Moore).

Functions of licks change during rainy seasons to those of natural water holes from which wildlife drank and wallowed. Some animals prefer using the drinking holes example are Buffalo and water-buck, but some still eat the soil lick mineral as well as drinking the lick example, warthog and hartebeest.

When salt lick appears, animals may travel to reach it, the lick become a sort of rally points where lots of wildlife can be observed. The concentration of animals in the area becomes visible, game viewing and photographing for tourist are improved in the area, animal studies and census conduction are also improved.

In an ecosystem, salt lick provides sodium calcium, iron, phosphorus and zinc required in the spring time for bones, muscles and growth for the wildlife. All trace elements like copper, magnesium and cobalt are retain in the salt lick for the metabolism of most mammals. Salt licks are used for hunting purposes by the animals.

Salt lick with a wide range of animals illustrate the ways in which wildlife naturally seek out nutrition which is essential to their survival. It provides nutrition for predicators in the form of conveniently located prey who may be distracted by the salt lick long enough to become a snack.

3. Species utilization of salt lick

The following species of wildlife usually utilized salt licks in Nigeria; Elephant (*Loxodonta africana*), Buffalo (*Syncerus caffer*), Western hartebeest (*Alcelaphus buselaphus*), Roan antelope

(*Hippotragus equimus*), water buck (*kobus defessa*), Kob (*kobus kob*), Bushbuck (*Traqelaphus scriptus*), Oribi (*Ourebi ourebi*), Red flanked duiker (*Cephalophus rufilatus*), Grimm's duikers (*Sylvicapra grimmia*), Warthog (*Phacochoerus aethiopicus*) *and* Baboon (*Papio Anubis*). The following game species were also observed in east Africa to use salt licks: rhino (*Diceros bicornis*) Zebra (*Equus burchelli*) Impala (*Aepycenus melampus*), Water buck (*Kobns ellipsiprymus*), Eland (*Taurotragus oryx*).

4. Activities of animals using salt lick

Salt licks are used predominantly in the day. Different species used salt licks at different period of the day. Buffalo and Warthog used lick mainly in the morning and afternoon. The shapes of the animal mouths determine the method by which salt licks material is obtained and ingested. Elephant and Warthog dig up the licks content using their tusks and lift up large licks materials with the trunk, whereas baboon used their hands to pick up and throw small pieces into the mouth. Hartbeest cut fresh lick materials by biting deeper into the craters with the incisors while water-buck, buffalo, roan, duiker, oribi and bush buck lick up the powdered lick material with the tongue.

Salt lick is found in abundance in the game reserves that are situated in the drier habitats in Northern Nigeria. Henshaw and Ayeni (1971) postulated that abundance and use of salt licks by wildlife indicates nutritional deficiencies caused by degrading environment or over-population or both. Examples are in the case of Yankari where ten licks are located within 11km along the Gaji River and the frequency of occurrence and intensity of use is very high.

5. Problems of salt licks in game reserved

Three problems have been noticed by the presence of salt lick in the game reserves; soil removal, vegetation destruction and spread of disease (Woodley, personal communication). Over 5000 tons of soils are removed annually. The soil is lost through wallowing, eating or licking and trampling. The areas around the salt lick are always trampled by hoof action, overgrazing and devoid of vegetation cover. Since many or so much drinking, urination, defecation, wallowing and feeding occur at the licks, diseases spread rapidly (Ayeni, 1979, Ogunsanmi, 1997). The use of artificial licks may prevent this.

6. Types of salt lick

Salt licks can be natural or artificial. Artificial salt licks are used by farmers for their cattle, horses and other herbivores to encourage health growth and development. Typically a salt lick in form of block is used in these circumstances. The block may be mounted on a platform so that domesticated animals do not consume dirt from the ground with necessary salt. There is need to medicate shy animals or a large group of animals.

Some people used artificial salt licks to attract wildlife such as deer and moose along with smaller creatures like squirrels. Animals may be attracted purely for pressure of humans who install the salt lick with the goal of watching or photography. Also salt licks are used by hunters to encourage potential prey to frequent an area. Wildlife biologists also used salt lick to assist them in tracking populations and can be a serve as medication in that it is used as birth control to keep animals from proliferating in the area where they are few natural predators, example is the deer.

Artificial salt lick comes in two forms; blocked and bagged. Bagged salt licks are designed to be buried in pits to create a more realistic form of salt lick with the salt and mineral leaching out in wet weather to form a salt deposit which will attract animals. While blocked licks are installed directly or mounted on platforms depending on personal taste. It can also be hanged on a tree in the middle of the farm or ranch house.

The universal popularity of salt licks with wide range of animals illustrates the way in which wildlife naturally seek out nutrition which is essential to their survival and provide nutrition for predators. Salt lick is a natural gathering place for grazing animals, which also attracts the carnivores. Animals ingest it as part of food they eat or they eat it directly

7. Significance of natural salt lick

Natural salt licks are utilizes by wildlife to supplement their mineral requirements (Ayeni, 1979). Wild game especially the herbivores can always identify the spots in their habitat where the essential minerals could be found. The African elephant have been noted to travel great distances to visit areas of saline earth which they swallow in considerable quantities as a purgative (Wari, 1993). Lick utilization is related to the spatial distribution and abundance of environmental minerals. A positive correlation between the spatial distribution of elephants in Wakie National Park, central Africa and the abundance of environmental sodium has been reported (Weir, 1972).

Water holes which usually serve as salt licks for animals during dry season may become heavily contaminated with infectious pathogens which can survive to the dry season (Woodford, 1979, Ogunsanmi, 1997). Droplet of respiratory disease is also made possible when animals crowd together in salt lick spots. Salt lick can be infected with anthrax spores and can act as focal point for the spread of the disease (Woodford, 1979). There is also a high increase in predation illustrated by the frequency of carcasses near licks which often lead to high mortalities (Lasan, 1999).

8. Resources availability and utilization by wildlife

The pattern of utilisation is determined by the growth of different plants and the physiological need of the animals, Benjamin, (2007). The abundant of mineral ions Na^+, K^+, Ca^+ and Mg^+ in salt lick, (Ayeni,1971), cause the concentration of big games along the river during the dry season due to the availability of salt lick, cover, water, and cover and fresh flush of the grasses and browse able materials.

Afolayan (1977), Milligant and Ajayi, (1978), observed that the utilization of salt lick by large mammals in Kainji Lake National park reduced with increased distance from water point particularly streams and rivers. Vertebrates have complex nutritional requirement in the form of chemical elements. Just as water and food supply, salt lick constitute one of the requirements expected in an ecological unit. Ayeni, (1971) observed that not all animals come regularly to salt lick but big games pay much visit e.g. elephants, antelopes, baboons etc. Phosphorus and sodium are believed to be principal trace elements causing animals to use salt lick (Cowan, 1949). Habitat degradation through over grazing and over browsing and soil compaction result from heavy lick use (Ayeni, 1972).

9. Factors influencing wildlife population

The increase in animal population is a function of animal chance to survive and multiply, and that chance to multiply and survive is a function of the environment. Environment is the sum total of all factors that influences the animals' speed of development and expectation of life and fecundity. The component which are claimed to be homogeneous with respect to the way it influences the animals chance of survival, Andrewartha and Birch, (1954) are;

- The resources e.g food, water and cover
- Mate
- Predators, pathogens and aggressors
- Weather condition
- Pollution.

10. Salt lick and trace minerals

Mineral elements play an important role in the nutrition of wild games. Hence a brief discussion on wildlife nutrition is important. Salt is unique in that animals have a much greater appetite for the sodium and chloride in the salt than for other minerals. Because most plants provide insufficient sodium for animals feeding and may lack adequate chloride content, salt supplementation is a critical part of a nutritionally balanced diet for animals. In addition, because animals have definite appetite for salt, it can be used as a delivery mechanism to ensure adequate intake of less palatable nutrient and as a feed limiter.

Sodium plays major roles in nerves and impulse transmission and the rhythmic of heart action. Efficient absorption of amino acids and monosaccharide from the small intestine requires adequate sodium. The other nutrient in salt, chloride, are essential for life. Chloride is a primary anion in blood and the movement of chloride in and out of the red blood cells, is essential in maintaining the acidic-base balance of the blood. Chloride is also a necessary part of the hydrochloric acid produced in the stomach which is required to digest most food. Animals have more defined appetite for sodium chloride than any other compound in nature except water. Ruminants have such a strong appetite for sodium that the exact location of salt source is permanently imprinted in their memory which they can then return to when they become deficient. Horses have been shown to have specific appetite for salt if the diet is deficient in sodium.

Trace elements

There are seven trace minerals that have been shown to be needed in supplementing animal diets. They are iron, copper, zinc, manganese, cobalt, iodine and selenium. They are needed in small amounts, or traces, in the diets and hence their names traces minerals".

Currently trace minerals deficiency is a bigger problem than the acute mineral deficiencies. They are several examples where an area was not recognise to have trace mineral deficient in the past but now has been shown to require supplementation. For example selenium deficiency was not considered to be a problem in the United States until recently.

Salt as a carrier of trace minerals

Salt is known to be a carrier of trace minerals, since all herbivores has natural appetite for salt this could serve as a source of trace minerals for them. Moreover when horses, cattle, sheep and other animals are on pasture with little, or no varying amounts of concentrate feeding, farmer can supply trace mineral salt in the form of a mineral block or loose trace mineral salt in a box.

11. Wildlife nutrition

Nutrition is the study of process by which organic and inorganic substances ingested by living organisms are converted to various needs for life processes such as promoting growth, replacing worn-out and injured tissues and perpetuate life. Wildlife nutrition is concerned with the supply of quality food in an animal environment. The basic requirements of all wildlife are food, water and cover. In general animal with adequate food supply grows larger, produced more young, are more vigorous and healthy, and are more resistant to many form of diseases than those affected by malnutrition (Nancy and Martha, 1995). When wild animals are shipped, moved, migrated, translocated or restocked to destinations where plant food that are typically consumed by freely grazing animals e.g reindeer are not found in sufficient quantities to provide adequate nourishment. Nutritional and digestive disorder set in which could be fatal (Luick, 1968). According to Luick, 1979, more than 60% of reindeer may die in two weeks of departure from their native tundra ranges.

Digestive and nutritional disorders are factors contributing to high mortality rate others include; the stresses of capture and handling, hyperthermia, regurgitation and inhalation of rumen contents, injury and diseases and overdoses of immobilizing agents and tranquilizers, (Luick, 1968, 1976). Nutritional disorders are primarily a result of failure in adjustment of the balance between nutrient input and requirements. These disorders are distinct and specific to a particular nutrient, (Sauvant, 1991)

An increasing amount of information is accumulating to show that many nutrients are needed at higher levels to improve the ability of the animal immune system to cope with infection. Sodium chloride, copper, zinc, Selenium, phosphorus and magnesium already have been shown to be helpful in this regard. Nutrients requirement for growth, gestation and lactation do not necessary mean that the level will be adequate for normal immunity and high resistance to diseases. But nutrient levels higher than those recommended may be needed for maximum productivity and health of the animals.

12. Essential mineral elements in wildlife nutrition

Mineral elements are restricted to mineral elements which have metabolic role in the animal body, (McDonald, 1979). Bowen, (1999), an essential mineral element is necessary to proof that diet lacking the element can cause deficiency symptom in animals. These depend on three basic factors. These are;

1. The organism can neither grow nor complete its life cycle without an adequate supply of the element.
2. The element cannot be replaced by any elements and
3. The element has direct influence on the organism and is involved in its metabolism.

13. Major mineral elements at salt-lick site

a. **CALCIUM (Ca):** Calcium is the most abundant element in the animal body. It is an important constituent of the bone and feet where about 99% of the body total calcium is found. Calcium is also an essential constituent of living cells and tissue fluids. It is essential in the activities of a number of enzymes including those necessary for the transmission of nerve impulses and for contractile properties of muscles. It is also necessary in the coagulation of blood and the normal function of cell membrane, (Clegg and Clegg, 1978; McDonald, 1998).

b. **PHOSPHORUS (P):** It occurs in the bone and is vital for bone formation. It also occurs in protein called phosphor-proteins, nucleic acids and phospholipids. The element plays a vital role in energy metabolism in the formation of sugar phosphate and adenosine di and tri-phosphates. It forms an essential constituent of milk and is necessary for the function of the neuromuscular system (McDonald, 1987).

c. **POTASSIUM (K):** Potassium plays an important role in osmoregulation of the body fluids and in acid-base balance in the animals. It functions principally as the cation of the cell. It plays an important part in muscle and nerve excitability, carbohydrate metabolism and is important blood and interstitial fluid, (McDonald et. al., 1981).

d. **CHLORINE (Cl):** Chlorine is associated with sodium and potassium in acid base balance and osmoregulation. It plays an important role in the gastric secretion where it occurs as hydrochloric acid as well as chloride salts. Chlorine deficiency can lead to an abnormal increase of the alkali reserve of the blood.

e. **SULPHUR (S):** It is an important constituent of protein in animals such as protein containing amino acids cystine, methionine, biotine and thiamine, the hormone, insuline and important metabolite, coenzyme A also contain sulphur.

f. **MAGNESIUM (Mg):** Magnesium is closely associated with calcium and phosphorus in the formation of bones and about 70% is found in the skeleton, the remains distributed in the soft tissue and fluids. It is the commonest enzyme activator, particularly in the activation of phosphate transferases, decarboxylases and acyl transferases (McDonald *et. al.*, 1987).

14. Trace elements

a.   **IRON (Fe):** Iron plays an important role in the synthesis of haemoglobin and enzymes in the foetus. It also occurs in the blood serum in a protein called transfrin which is concerned with the transport of iron from one part of the body to another. Iron is also a component of many enzymes including the cytochrome C reductase, succinic dehydrogenase and furmaric dehydrogenase. Hence, it is vitally important in the oxidative mechanism of all living cells. (Clegg and Clegg, 1978; Pethes, 1980; McDonald, 1987). Symptoms of deficient of iron include paleness of the skin, hypochromic microcyclic anaemia and conjunctiva.

b.   **COPPER (Cu):** Copper is found in blood plasma and occurs in various complex forms loosely bound to albumin and amino acids. It is necessary for heamoglobin synthesis and has been found in over 35 enzymes and proteins. These include; red blood cell, cerebrocuprein, and mitochondrocuprein, (Underwood, 1977; Pethes, 1980).

c.   **IODINE (I):** Iodine is required by animals for the synthesis of the thyroid hormones, thyroxin and triiodothyroxin produced in the thyroid gland (Pethes, 1980). Deficiency results in endemic goitre.

d.   **ZINC (Z):** Zinc is found in every tissue in the animal's body. It plays an important role in enzyme activation and in wound healing. It is important in the fundamental process of RNA, protein synthesis and metabolism. Deficiency results in growth retardation, loss of apetite, alopecia, bone deformation, impaired fertility in both male and female animals and increase occurrence of tetratogenicity and behavioural anomalies.

e.   **FLOURINE (F):** It is one of the constituents of bone, teeth, soft tissues and body fluids. It activate adrenal cylase enzyme which is the primary mediator of hormone action (Pethes, 1980). It's function in the prevention of dental canies. Deficiency symptoms include; pain on movement, lameness, arthritis, of the hip, erosion of the tooth enamel and decrease in milk yield, (Pethes, 1980).

15. Study area

The study was conducted in Kainji Lake National Park (KLNP). In Nigeria the role of game reserve in conserving wildlife for various purposes is widely recognised. The flora (plants) constitute only one element of the complex ecosystem which they belong and are not in stable state. The effective management of land as game reserve is the general principle. There is a management plans to be drown. Such plan has been prepared for Kainji Lake National Park (KLNP) (Ayayi and Hall, 1975). The principle purpose of the plan is to provide all available information relevant to the management of the reserve. It also makes provisions for it regular revision and updating and incorporate timetable for these purposes.

16. Location

Kainji Lake National Park extends 80km in an east-west direction and about 60km north-south. It consists of two sectors, the Borgu sector and Zuguruma sector. It lies between $90^0 4^1$ to $10^0 30^1$ N and $3^0 30^1$ to $5^0 50^1$ E, covering a total area of 5,340.82km (Tuna, 1983). The Borgu sectors cover an area of 3,970.02km S.E in Borgu Local Government Area of Niger State. Ero, (1985), put the location of Borgu between $10^0 50^1$ N latitude and $4^0 19^1$ E longitude. The Borgu has the Kainji Lake on the east while the west is by the republic of Benin (Figure 1).

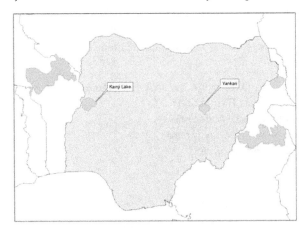

Figure 1. Map of Nigeria showing the location of Kainji Lake National park

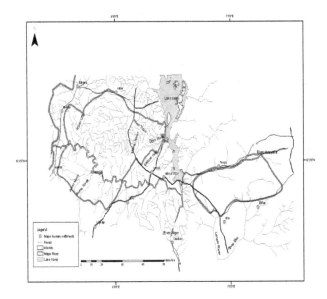

Figure 2. Showing the Geo-reference Map of Kainji Lake National Park in respect to the Lake and Salt lick sites.

17. Vegetation of the study area

The vegetation of Nigeria vegetation consists of forest, savannah and Montane. The forest zone comprises mangrove forest, rainforest and dry forest southern and northern guinea savannah zone, Sudan zone and Sahel constitute the savannah vegetation. Montane vegetation is formed by the Montane forest and grassland. The derived savannah forms the boundary between the forest and the true savannah vegetation types.

The Borgu sector of Kainji Lake National park has transitional vegetation which is between the Sudan and the Northern Guinea savannah type (Onyeanusi, 1996). FAO, (1974) also included the vegetation of the study area in the guinea zone in the map of vegetation of Nigeria. FAO, (1974) divides the vegetation into six main vegetations having significance as wildlife habitat types. These are:

a.   The Burkea/Detarium microcarpum (Wooded Savannah)
b.   The Isoberlinia tomentosa (Woodland)
c.   Diospyros mespilliformis (Dry Forest)
d.   Terminalia macroptera (Tree Savannah)
e.   **The Riparian forest and woodland**
f.   **The Oli complex**

Burkea africana/Deterium microcarpum (Wooded savannah)

This consists of the majority of the vegetation of study area. The density and height of woody cover varies with soil and influence of fire. *Afzelia africana* is an associated species which occasionally form patches of woodland. Trees and shrubs common in this area are *Butyrospermum paradoxum, Terminalia avicenniodes, Parinari polyandra, piliostgma thonnigii, Maytenus senegalensis, Gardenia ternifolia, strychnos inocua* (FAO, 1974).

Some of the grasses associated with this community include *Hyperrhenia involucrata, Andropogon gayanus, A. pseudapricus, Genium newtonii, Chlospermum tinctorum, Indigofera bractolats, Hyperthelia dissolute, Brachaiaria jubata, B. brachylopha* and *Aristida kerstingii.* (FAO1974).

Isoberlinia tomentosa (Woodland)

This associates with quartzite ridges and are extensive on higher ground in the study area. The *I. tomentosa* occur in almost pure stands though occasionally *I. doka* is found where stone intrusions occur. *Afzelia africana* and *Ostryoderris stuhlmanni* is found adjacent to *I. tometosa* woodland. Grasses such as *Monocynibium cresiforme, Schizachhyrium sanguineum, Beckeropsis uniseta, andropogon tectorum* and *Andropogon gayanus* occurs extensively on poorer sites.

Diospyros mespiliformis (Dry forest)

This distinctive type occurs as units of a few acres extent at scattered localities in the central part ot the study area. *Diospyros mespiliformis* foms the bulk of the tree layer while

Polysphaeria arbuscula usually comprises of a dense under storey. The grasses of this type are the broadleafed *Opilsmenus hirtellus* (FAO, 1974).

Terminalia macroptera (Tree savannah)

This occurs in the low- lying seasonally inundated areas, characterised by dark grey hot-wallowed clays. Typically there are few woody plants in this type. They include *Pseudocedrela kotschyi, Mitrangyna intermis* and *Daniella oliveri* (FAO, 1974). The grasses are those associated with swampy ground. They include *Panicum paucinode, Brachiara jubata, Hyparrhenia glabriuscula, H. rufa, Andropogon perligulatus, A.pseudapricus, Schizachrium schweinfurhii,Hyparrhenia cyanescens* and *Echinocloa obtussiflora* (FAO,1974).

Riparian forest and woodland

This includes the vegetation of all water courses of the reserve with the exception of the Oli and the lower reaches of its two largest tributaries- the Uffa and the Nanu. The riparian/forest woodland which develops in response to the prevailing high atmospheric water content, resembles the moist savannah bordering the forest zone. The common tree species to this vegetation type include: *Cola laurifolia, Irvingia smithii, Antidesma venosum, Pterocarpus santalinoides, Diospyros mespiliformis, Daniella Oliveni, Gardenia Species, Strychnos spinosa, Terminalia aricennoides* and *Maytenus senegalensis*. The grasses remain green into greater part of the dry season. The grasses include *Acroceras zizanoides*.

The Oli complex

The Oli complex is distinctive in its heterogeneity. It is heterogeneous vegetation which characterises the courses of the Oli River as well as some parts of its main tributaries (e.g. River Nanu, Suna and Suma). Its dominant tree species are assemblages of characteristics rainforest species e.g. *Annogeissus leioxarpus, Vitex doniana, Khaya senegalensis, Mitragyna inernus, Chlorophaora excels*. The oli complex and the riparian forest are closely associated.

The valleys have a scattering of *Terminalia macroptera* in grassland of lots of *Hyperrhenia smithiana*. The upland areas are very distinctive but small unit of *Diospyros ruspiliformis* forest. The evergreen nature of the *Diospyros ruspiliformis* and the shrub makes it difficult for fire to penetrate it. *Oplismenus hitellus* grass also occurs in this unit. In the upland, three woodland units are found they are, *Isoberlinia doka, Isoberlinia tomentosa* and *monotes kerstingii* woodlands. The largest vegetation is covered by *Burkea Africana-Detarium macrocarpum* wooded savannah.

18. Climate of Kainji Lake National Park

Rainfall: The two major features of the climate of the park are the wet and dry season and they are variable from year to year. The wet season extends from May to November while the dry season extends from December to April. The highest amount of rainfall is always in

	2007	2006	2005	2004	2003	2002	2001	2000
JAN	-	-	-	0.2	-	-	-	-
FEB	-	-	-	-	2.6	-	-	-
MAR	-	-	28	2.8	18.7	-	-	-
APRIL	65.51	30	53.3	67.4	45	84.9	48.5	8.6
MAY	100.7	168.4	58.9	167.6	82.2	34.68	96	70.43
JUNE	119	225.7	179.04	176.5	203.27	103.62	229.9	227.4
JULY	175.8	226.33	184.6	231.3	111.3	154.15	165.23	115.1
AUG	306	166.8	106	158.3	126.56	159.08	111.21	253.8
SEPT	127.2	255.06	329.39	265.76	147.17	122.9	211.65	245.6
OCT	15.9	69.3	110.4	112	110.3	74	119.98	50.12
NOV	-	-	-	12	-	3	-	-
DEC	-	-	-	-	-	-	-	-

Table 1. Kainji Rainfall Data from 2000-2007, Kainji monthly Rainfall (mm)

August (11.89mm) while the lowest rainfall (2.090mm) is in October (Table 1). These values varied yearly. Milligan (1979) indicates that there is a decreased in rainfall from the south to the north and increased rainfall towards the west and east and low condition in central northern regions.

Temperature: The highest is in the dry season just before the rain and lower during the wet season it picks up again towards the end of the wet season and later drops to the lowest value in December and January during harmattan. The minimum temperature during this period ranges from 17.87°C to 19.90°C (Onyeanusi 1998).

Relative humidity: The relative humidity increases gradually from low values at the beginning of the dry season to a peak during the wet season. Generally, the relative humidity follows opposite pattern to that of temperature.

Wind: This influences both incidence and duration of the wet season considering the whole year, southern winds predominate over northern winds in the Borgu sector of the park. There is also a distinct seasonal trend, with the dry, dusty, northern winds prevailing during the beginning of the dry season that is November to February, while the moist southern winds prevail throughout the wet season.

Topography: The landscape of the Borgu Sector of the Kainji Lake National Park is gently undulating. Its features may be related to the lithology and erosion history of the area. The relief is broken in places by quartzite ridges (FAO, 1974). Elevation of the central and Western parts of the park lie between 800 and 1,000ft the highest point in the park is 1135ft in the Northwest. Kubil hill, just outside the Northern boundary of the park has an elevation of 1,684ft (FAO, 1974). The land slope down from the East to the Niger valley. The lowest elevations are along the Kainji shore where the normal maximum high water mark is 465ft contour (FAO, 1974; Afolayan, 1977).

Geology: The Borgu sector of the park is underlain by the Basement Complex helica was considered to be Pre-Cambrian in origin (Afolayan, 1977). Most of the area is composed of geissose rock and other units consist of younger metasediments which are mainly schists and phyllites. According to FAO, (1974), the park is underlain by undifferentiated metasediments in the east and west. The Basement Complex was until recently considered as pre-cambrian. However, Truswell and Cope, (1963) presented age determination which will place it in late Pre-cambrian or early Paleozoic.

Soil: Detailed soil survey of the Borgu sector shows that the soils in the area are low of fertility. It is slightly acid to natural and the acidity increases irregularly with dept. Although the soil nutrient is low, well developed and maintained savanna woodland exists on this soil. Meanwhile, the park has over ferruginous tropical soil and crystalline acid rocks (Anon 1964). Also the soils are shallow.

Drainage: The drainage is to the east because of the slope into River Niger through River Oli which is the largest river in the park. The river Oli and its tributaries drain the western part of the Borgu sector while the eastern part are drained by rivers Doro, Timo and Menai into Kainji lake. The river covers an estimate of about 3.305kms from the Nigeria border with Republic of Benin to where it empties into the Niger River. The river has maximum flow of approximately 600-700mm^3/sec of water at the end of wet season breaks into pools in the dry season. The main tributaries of the Oli river are; Uffer, Koa, Nanu, Suma and Suna. Though other unnamed seasonal rivers also feed the Oli river in the wet season. There are six drainage basins in the western part of Borgu sector, viz the lower and upper Oli, Suna, Nanu, Uffa, and an unnamed basin which covers about three quarters of the sector. In the eastern part are seven drainage basins giving a total of thirteen basins. Eroded water from these rivers contributes to the volume and rate of flow of the Oli river in the early part of the wet season. During dry season, surface flow ceases in all the major rivers particularly the Oli River and only pools remain which provides sources of perennial water for wildlife population.

19. Research methodology

Reconnaissance survey: The reconnaissance survey was carried out in the month of June 2008 to get acquainted with the terrain of the park. Information from the park management and the review of previous literature on salt lick mineral in the park served as a bedrock towards successful reconnaissance survey. This aided in choosing the study side and the ecological survey method used. The study covers between June 2008-February 2009 (6month). The following were predetermined to aid sampling design.

Ecological survey: Based on the methodology adopted from Ayodele,(1988), Lameed,(1995) and Akanbi,(1997), which stated that an ecological survey for an area should be conducted on a comparative bases, particularly the heterogeneous to indicate a long term range. The study was carried out in the Borgu sector of Kainji Lake National Park. Out of twenty salt lick areas in the Kainji Lake National Park, four were selected based on concentration of the salt and species to the spot, management and tourists' preference and proximity to the camp. The inventory of the place was taken which are:

a. Inventory of the salt lick in three places, that is, the middle, the upper and the lower parts to look at the composition of the minerals.
b. The fauna and vegetation were assessed by laying four (4) transect 1km each in each of the salt lick. This transect was taken towards the north, south, east and west which form the radii of the utilized areas.

Point Centre Quarter Method (Vegetation sampling): For vegetation, point centred quarter method were taken at every 10m and the fauna were assessed looking into the direct and indirect survey of fauna species to note the absolute and relative density of the species around the area. The relative density is to asses the significant of the spur rate of the species along transect by identifying the faecal samples, foot prints, death/ carcass and calls of the species available within the transect (Hopkin, 1974).

A transect of 1km was cut from each of the sites with the use of a cutlass and a compass to align the transect line. Each transect was marked at 50m intervals. Daily survey of the transect along the salt lick was carried out as early as 6.30am with the assistance of the research officer and a ranger from the park and backed by 2.pm. The early morning study was designed to survey some of the wild animals that are inactive and sluggish in the afternoon whose detection may be difficult and liable to error. At the beginning of each early outing on the salt lick, the following were recorded;

- Location and the name of the tract
- Date
- Time (beginning and stop)
- Number of the salt lick.

Line Transect Survey: A line transect according to Roger, (1975), in a fixed path independent of external features along which survey will take place. At each outing the following equipments were used;

- Binocular for clearer viewing of animals at a long distance
- Camera for taking photographs
- Field notebook, pen and pencil
- A cutlass
- A wristwatch.

The following information was recorded at each salt lick sites

- Name, number and species of the flora and fauna around the salt lick sites
- Activities of the animals
- Droppings of the named and species of the animal found
- Footprints of the named animals
- Food materials. Carcass of the animal
- Calls of the animal

-     Habitat type

20. Search method for direct and indirect methods

Both the direct and the indirect methods were used to study the animals that visit the salt lick. This involved walking along the delineated transect line and looking for signs of species presence. The stop and search method was used as an indirect method of estimating wildlife population through the use of faecal droppings, footprints, feeding/remnant and calls of the animals. The following assumptions for animals that visit salt lick was outlined by Burham et al 1980; Seber, (1982). These are

1.   The number of footprints of a particular animal was closely examined and traced to the salt lick.
2.   The number of faecal samples was observed along the transect
3.   The feed/remnants of animal along the transect were also examined
4.   The activities of the animals around the area and along the transect was examined
5.   Sighting of animals was carried out
6.   The faecal droppings containing salt lick was picked and examined
7.   Backward movement along transect to confirmed the samples was limited and observation period not more than 10minutes.
8.   Samples positioned directly over the transect line was not missed. It was also understood that not all the sample within the survey area was detected. Some was inevitably missed and possibility of detection declined with increase distance from the transect line or survey path.

21. Population density

To determine the population density for each species of the animal encountered during the survey. Time species count was used as outlined by (Ajayi, 2001). The formula is as follows

$$P = \frac{AZ}{2XY}$$

Where:
A = Area occupied by the species
Z = number of animal seen
2 is constant
X = sighting distance
Y = length of transect

The area (A) is determined by multiplying the transect width by the length of the transect.

The number of the animals seen (Z) = species seen/transect

22. Data analysis

All data collected were subjected to appropriate statistical analysis depending on the nature of the study. Correlation, Analysis of variance, descriptive analysis and T-test to differentiate two variables were used to draw conclusion.

Graphs, photographs and diagrams were used at appropriate places for proper result. Chi-square test was also used to analyse the result.

23. Results and discussion

	A	B	C	D
Kobus kobs	65	51	20	20
Hippotamus	39	21	13	33
Roan antelops	1	10	4	-
Red flank dunker	5	13	1	-
Hares	1	2	-	-
Baboons	10	13	9	4
Silvet cat	2	3	5	2
Bush buck	9	2	-	-
Monkey	4	3	2	1
Lion	1	5	-	1
Flankolin	-	-	-	1
Crocodile	-	-	-	1
Total	137	123	54	63
Average	13.70	12.30	7.71	7.86

Table 2. Table 2: Diversity of Wild animals sampled within the various transects studied.

Table 3 shows record of animal sample within various transects at the four sites (A, B, C and D). It was observed that the highest number of animals is recorded in transect A (137) with an average number of 13.7, followed by transect B (123) with an average number of 12.3. the lowest value is recorded in transect C (54) with an average of 7.7.

Parameters	df	ms	f	p-level
Salt lick1	3	81687.34	10.1038	0.0013+
Error	12	8084.833		
Total	15			

Table 3. ANOVAs for Number of trees/hectare

The table 3 above shows that at p-level there was no significant difference (p>0.05) in the number of trees/hectare. Therefore the alternative hypothesis was accepted that said there was no significant difference in the trees/hectare in the salt lick sites.

Parameters	df	ms	f-cal	p-level
Salt lick2	3	305.58	2.0204	0.1648
Error	12	151.25		
Total	15			

Table 4. ANOVAs for number of different trees at the salt lick sites (tree diversity)

Table 4 above shows that there was significant difference (p<0.05) in the tree diversity across transects at the salt lick sites. Therefore the null hypothesis is accepted while the alternative is rejected.

Parameters	df	ms	f-cal	p-level
Salt lick3	3	1757.67	.390	.7623
Error	12	4702.5		
Total	15			

Table 5. ANOVAs for number of animals/hectare at salt lick sites

The table 5 above shows that at (p<0.05) there was significant difference in the animals across the four transects. Therefore a null hypothesis was rejected while the alternative hypothesis was accepted.

Parameters	df	ms	f-cal	p-level
Salt lick4	3	39.3333	.39008	.762310
Error	12	100.8333		
Total	15			

Table 6. Anova for different animals/hectare at the salt lick sites (animal diversity)

The results obtained from the direct and indirect survey shows that f-cal was significantly difference at (p>0.05). Alternative hypothesis was accepted that said that there is significant difference in animal species across transects as seen in table 6 above.

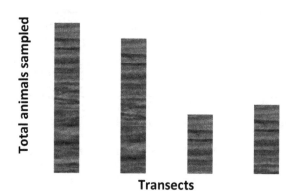

Figure 3. Variation of animal sampling within transects studied.

Figure 3 above and the graph sows that the highest animal sampling was recorded in transect one (13.7), followed by transect two (12.3) and the lowest value was in transect three (7.7).

The table 7 below shows the records of various tree sampled within the transects. Transect three(C) recorded the highest value of trees (213) and the average is 14.20 while the lowest is recorded in transect A (149) and the average of 9.93. Meanwhile *Acasia spp* was found to be dominant species within the four transects.

	Transects			
Trees	A	B	C	D
Acasia spp	42	84	28	81
Annogeissus leiocarpus	30	11	33	6
Daniella Oliverie	6	22	6	14
Terminalia spp	32	8	37	10
Detarium microcarpum	4	4	51	2
Burkea Africana	12	7	12	4
Gardenia sokotolensis	2	7	1	1
G. aquala	8	6	9	6
Strychnos spinosa	1	1	4	14
Bytyrospermum paradoxum	1	3	6	9
Darsperus mispiliformis	1	4	1	1
Tamarindus indica	5	7	3	2
Afzelia Africana	2	3	5	2
Tetrap tetrap	2	4	12	6
Acaria spp	1	4	5	2
Total	149	175	213	160
Average	9.93	11.67	14.20	10.67

Table 7. Diversity of Vegetation sampled within transects studied.

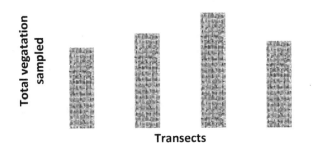

Figure 4. Variation in total vegetation (trees) samples within transects.

Figure 4 above shows that transect three has the highest tree samples (14.2) followed by transect two (11.67) and the lowest was recorded within transect one (9.93).

The results obtained from the direct and indirect survey shows that f-cal from the two survey types was significantly different ($p > 0.05$) fro each other. Alternative hypothesis was accepted that said there are significant differences in the trees sampled as seen in table 8 above.

ANOVA: Single Factor

SUMMARY

Groups	Count	Sum	Average	Variance
A	15	149	9.933333	179.2095
B	15	175	11.66667	424.9524
C	15	213	14.2	238.3143
D	15	160	10.66667	397.8095

ANOVA

Source of Variation	SS	Df	MS	F	P-value	F crit
Between Groups	156.1833	3	52.06111	0.1679	0.917595	2.769431
Within Groups	17364	56	310.0714			
Total	17520.18	59				

Table 8. ANOVA of trees sampled with the transects groups.

Salt licks	%Na	%Ca	%Mg	%K	Fe (mg/Kg)	%PO4	%SO4
1	3.69	1.79	0.88	6.76	1149	0.73	0.35
2	4.18	3.54	1.23	4.46	992.81	1.93	0.55
3	3.79	1.92	0.81	6.81	1168	0.89	0.39
4	4.21	1.85	0.79	6.95	1173	0.87	0.43

Table 9. Chemical analysis of mineral contents in salt licks.

Fig.5 shows that the level of sodium (Na) is highest in salt lick site four (4.21%), followed by salt lick site two (4.18%). There is no much different between the percentage level of sodium salt lick sites one and three (3.695 and 3.765), but the lowest was recorded in salt lick site one.

Salt lick site two has the highest percentage level of calcium (3.54%), while the lowest percentage level of calcium was recorded in salt lick site one (1.79%). Salt lick site three and four has (1.92% and 1.85%) respectively (Fig. 6).

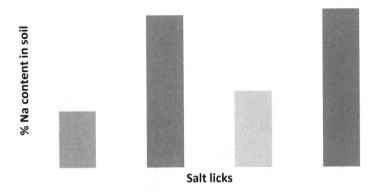

Figure 5. Concentration of Na (%) in salt licks studied.

Figure 6. Concentration of Ca (%) in salt licks studied.

Figure 7. Concentration of Mg (%) in salt licks studied.

Fig.7 above shows that the salt lick site two has the highest level of magnesium (1.23%), followed by salt lick site one (0.88%). The lowest value was recorded in salt lick site four (0.79%).

Figure 8. Concentration of K (%) in salt licks studied.

Among all the minerals present in the salt lick sites, potassium was found to be the highest mineral content. Fig. 8 above shows that salt lick site four has the highest level of potassium content (6.95%), followed by salt lick three(6.8%), salt lick one(6.76%) and salt lick two has the lowest (4.46%).

Figure 9. Concentration of PO₄ (%) in salt licks studied.

Figure 9 shows the highest percentage level of phosphate in salt lick two (1.93%). The lowest level is in salt lick one (0.73%). Salt lick three and four have (0.89% and 0.87%) respectively.

Iron is measured in milligramme per kilogramme (mg/kg) and not in percentage. The highest value is in salt lick four (1173mg/kg), while the lowest is in salt lick two (992.81mg/kg). Salt licks one and three have (1149mg/kg and 1168mg/kg) respectively (Fig. 10).

Figure 10. Concentration of Fe (mg/Kg) in salt licks studied.

Figure 11. Concentration of SO₄ (%) in salt licks studied.

Sulphate was found to have the highest percentage value in salt lick two (0.55%), and the lowest percentage value in salt lick one (0.35%). Salt lick four has the value of (0.43%) and salt lick three has 0.39% (Fig. 11).

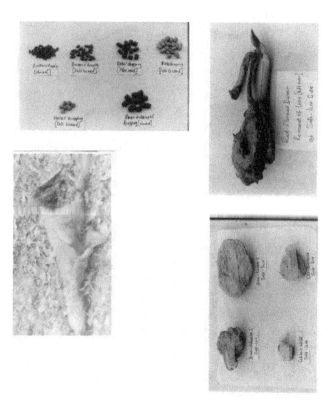

Figure 12. Droppings of wild animals and remains of some species killed by carnivores utilizing the saltlick site within the KLNP.

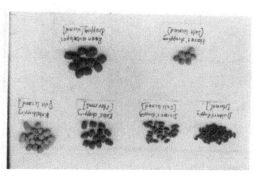

Figure 13. One of the carcass of infant herbivores found at the saltlick site with the collected droppings.

Figure 14. Elephant is one of the common preferred species visiting saltlick site in the Park.

24. Conclusion and recommendations

Kainji Lake National Park is an important reservoir for several wild herbivores. From this study, the results show that the highest population of animal was recorded in salt lick site three (3) and one (1), that is the Oli Entrance and kilometre three Drum. This was significantly different ($p < 0.05$) and higher than those in salt licks four (4) and two (2). Therefore it could be deduced that herbivores have more preference to salt licks near river sides than those that are not.

The population density of the animal in the area was established by indirect/relative survey, since it has to do with the observation of some indices like footprints, feed remnants/signs, and faecal droppings. Carnivores were also noticed at the study site like lion by the carcases and call of the animal. It was observed that there was an interrelationship between animals in the lick as some fruits were found at the lick sites. Most of the animals were not seen due to their nocturnal nature and shyness. Therefore indirect/relative method of transects survey was ideal to established their presence, distribution and abundance. The activities of the animals were also assessed.

From the study, the result of the analysis shows that both the macro and trace elements are present in the salt licks of the park. The mineral elements identified in the salt lick are; calcium, sodium, potassium magnesium, phosphate, iron, and sulphate. Potassium is most abundance mineral element in the salt licks while sulphate is the least abundant. The analysis revealed that the salt lick of the Oli complex contain the highest potassium and iron. This is probably due to the accumulation of dead leaves, defecations, urination and soil parent materials. The salt lick one has low calcium and sodium mineral elements due to the fact that there was less abundance of vegetation due to erosion of the upper slope of the reserve. Hence these minerals could have been leached away by erosion.

Salt lick spot are quite abundance in Kainji Lake National Park (KLNP) which is frequently used by different herbivores and carnivores. Hence the lick plays an important role in supplementing minerals lacking in the animal's diets. The licks therefore helps in preventing nutritional diseases and disorders that could result due to lack of these minerals and also in biodiversity conservation of wild fauna species of the park.

25. Recommendations

The results show that different herbivores utilize the salt licks in the park and this cannot be completely exploited on short duration and single handed project of this nature. It required continuous research work to be carried out on many other areas of interrelationship with natural ecosystem where they are found. Effort should be made to establish the proximate population and consistence to monitor the population. This will enable the management of the national park to know if the government and non governmental investment is achieving the desired impact.

Efforts should be made towards the rate of poaching activities in the park by reintroducing the special squad to complement the effort of the forest guards. Communication equipments like walkie-talkie, GSM antenna should be provided in the park, handsets, boots and patrol clots should be provided for the guards to carry out their duty effectively. The numbers of park guards present are grossly not adequate as compared to the increase number of poachers daily. More park stations should be created to boost the morals of the guards and more sophisticated arms given in view of the risky nature of the job. Residential quarters should be built by government for the park guards to live as the situation of rented building is exposing them to high risk and at the same time stand a chance of compromising their work with the poachers. Good incentives like food, medicals allowances should be provided for the park guards. Unannounced visit to the park station by senior officers should be vigorously carried out to ascertain if guards are on patrol regularly.

The Support Zone Development Programme (SZDP) should supervise the disbursement of loans to the community and the use of the loan to the community. The enlightenment on the conservation should be fully understood by the people. Therefore it should go beyond radio, television announcements, pamphlets and seminars, but house -to- house enlightenment should be used for the message to be fully understood. Infrastructures should be provided for the host community to justify the forfeiting of their resources to the government. Indigenes should be given scholarships to enable them go to school to reduce their dependence on hunting and logging which they carried out as a means of livelihood.

Author details

A. G. Lameed
Department of Wildlife and Ecotourism Management, Faculty of Agriculture and Forestry, University of Ibadan, Ibadan, Nigeria

Jenyo-Oni Adetola
Department of Aquaculture and Fisheries Management, Faculty of Agriculture and Forestry, University of Ibadan, Ibadan, Nigeria

26. References

Afolayan, T. A. (1977): Savannah Structure and Productivity in Relation to Burning And Grazing Regime in KLNP, unpublished PhD Thesis. Dept. of Forest Resources Management. University of Ibadan.

Ajayi, S. S. (1979): Utilization of Forest by Wildlife in Africa, F. A. O. Rome 1979

Akanbi, O. A. (1997): An Ecological Basis For Management of Enamania Game Reserve in Benue State, Nigeria. Ph D. Thesis unpublished, Department of Wildlife and Fisheries, University of Ibadan.

Alloway, B. J. (1990): Heavy Metals in Soil. John Wiley and Sons Inc. New York.

Andrewartha, H. G. and Birch, C. C. (1954): The Distribution of Abundance of Animal in Chicago, University of Chicago Press.

Andrewartha, M. C. (1954): Animal Environment and Man in Africa. 3rd Edit. 1974 Pp 120 – 134.

Anon, O. (1991): The Role of Tropical Timber Extraction in Species Extinction. *London Friends of the Earth.*

Ayeni, J. S. O. (1971): Some Aspect of Natural Mineral Licks by Large mammals of Yankari Game Reserve, North Eastern State of Nigeria. B.Sc Desertation in Forest Resources Management. University of Ibadan . Vol. 7 pp 47-53.

Ayeni, S.O. (1972): Chemical Analysis of some soil samples from Natural Licks. Vol, 2 pp 16-23.

Ayeni, S. O. (1975): Utilization if water holes in Natural Parks. *East Africa Wildlife Journals.* Vol. 13 pp 305-323.

Ayeni, S. O. (1979): Utilization of Mineral salt Licks. Wildlife Management in Savana Woodland. Cambridge University Press.

Ayodele, I. A. (1988): An Ecological Basis for Management of Old Oyo National Park, Unpublished Thesis Department of Wildlife and Fisheries Management University of Ibadan.

Benjamin, B. I. (2007): Preservation of Africa's Biodiversity. In Nature Fauna FAO/UNDP publication, Vol. 8, No 1. 1992.

Bowen, H. D. M. (1979): Environmental Chemistry of Mineral Elements. Accademic Press, London.

Buckman, H.O. and Brandy. N. C. (1960): Reaction of Saline and Alkaline Soil. 6th Edition pp 567.

Budd, J. (1973): Infectious Disease of Wild Mammals. (Ed. J. W. Davis, L. H. Karstad and D. O. Trainer). The Iowa State University Press, 36-49pp.

Burnham, A. et al (1980): Estimation of Density from line transect Sampling of Biological population of Wildlife Monogr. 72: pp 202.

Clegg, P. S. and Clegg, A. G. (1978): Biology of Mammals. Williams Heineman Medical Book Limited, London.

Cowam, I. McTaggart and Brink. V.C. (1949): Natural Game Lick in the rocky Mountains National Parks of Canada. Jour of Mammalogy, vol 30 pp 387.

FAO, (1983): Kainji Lake Research Project. An Ecological Survey of Borgu Game Reserve, UNDP/FAO Tech. Report, Rome.

Friend, M. and Trainer, D. O. (1970): Serology of Newcastle Disease in Captured Mallard and Swams. Jour. on Wild Diseases.Vol. 6 pp130-137.

Heady, H.F. and Heady, F.B (1982): Range and Wildlife Management in the Tropics Vol 2. Longman Pub. New York.

Heady, H. F. and Child, R. D. (1994): Rangeland Ecology and Management. Westview Press.

Hedrick, D. W. (1966): What is Range Management? Pp 19-11.

Lasan, A. H. (1979): Habitat Management for Wildlife Conservation with Respect to Fire, Salt Lick and Water Regime. Management in Savannah Woodland (Ed. S. S. Ajayi, and L. B. Halstead). University Press.

Lameed G. A. (1995): Some Aspect of Ecology of Wildlife in Gashaka Gumti National Park. Unpublished M.Sc. Thesis University of Ibadan.

Luick, J.R. (1979): Nutrition and Metabolism of Reindeer Caribou in Alaska with Special Interest in Nutritional and Environmental Adaptation. Energy Commission 1969, pp 1-32.

Mertz, W. (1976): Trace Elements in Animal Nutrition. Prog. of Symp-Nuclear Techniques in Animal Production and Health, FAO/IAEA, Vienna.

Milligan, K. R. N. (1979): An ecological basis for the management of Kainji Lake National Park, Nigeria. Ph D Thesis. University of Ibadan.

Nancy, A. S. and Haufler, A. A. (1975): Mineral Elements in Animal Nutrition. *Jour. in Animal Nutrition.* 36(5): 169.

NEST, (1991): Nigeria;s Threatened Environment; A National Profile , Nigeria Environmental Action Team, Publication. Ibadan. Pp 1-28.

Okeyoyin, O.A. (1988): Behavioral Ecology of Wildlife in KLNP. A Study of food resources exploitation. Unpublished M. Sc Thesis. Dept of Wildlife and Fisheries Management. University of Ibadan.

Popoola, L. (2000): Practise of Environmental Management. *Forest Production Paper* Presented at International Enabling Conference. Abuja. 9th-10th oct, 2000. Pp 26.

Pethes, G. (1980): The Need for Trace Element Analysis in the Life Sciences. Elemental Analysis of Biological Materials. IAEA, No 197, Vienna.

Roger, O.U. (1975): Ground Census Survey of Wildlife in Woodland Habitat.

Seber, G. A. F. (1982): Estimation of Animal Abundance and Related Parameters 2nd edition, Macmillan. New York.

Society of Range Management (1989): Terms used in Range Management. PP 39.

Stoddart, L.A. (1967): Range Management Techniques. Pp 20-30, 37.

Underwood, E.J. (1977): Trace Elements in Human and Animal Nutrition. Press. New York.

Wari, M. (1993): Know the Natural History of Nigeria Wildlife. Park News, 1(5) 12-15.

World Book Encyclopaedia (1995): Vol. 19. World Book International London.

Chapter 5

Characterization and Biological Activity of *Bacillus thuringiensis* Isolates that Are Potentially Useful in Insect Pest Control

Analía Alvarez and Flavia del Valle Loto

Additional information is available at the end of the chapter

1. Introduction

Bacillus thuringiensis (*Bt*) is a spore-forming bacterium well-known for its insecticidal properties associated with its ability to produce crystal inclusions during sporulation. These inclusions are proteins encoded by *cry* genes and have shown to be toxic to a variety of insects and other organisms like nematodes and protozoa [1]. The primary action of Cry proteins is to lyse midgut epithelial cells through insertion into the target membrane and form pores [2]. Once ingested, crystals are solubilized in the alkaline environment of midgut lumen and activated by host proteases [3]. On the other hand, the involvement of *Bt* proteases in processing inactive protoxins is also reported [3]. These toxins are also highly specific and completely biodegradable, hence no toxic products are accumulated in the environment. In fact, Calderón et al. [4] suggest the potential use of some crystal proteins as adjuvants for the administration of heterologous antigens.The activity spectrum of *Bt* toxins continually increases as result of the ongoing isolation of new strains around the world.

The fall armyworm, *Spodoptera frugiperda* (*S. frugiperda*) (Lepidoptera: Noctuidae), and the variegated cutworm, *Peridroma saucia* (*P. saucia*) (Lepidoptera: Noctuidae), are two lepidopteran pests that cause severe damage to a variety of crops. While the first one mainly attacks corn, rice, peanuts, cotton, soybeans, alfalfa and forage grasses [5], the second one targets peanuts, sunflowers, soybeans and grapevines, among others [6]. Currently, control of this pest relies on chemical insecticides. Nevertheless, the rapid increases in resistance to insecticides together with the potential adverse environmental effects produced by these chemicals have encouraged the development of alternative methods for Lepidoptera control [7,8]. Among these methods the use of *Bt* as a biocontrol agent has shown to be extremely valuable. The diversity of Cry toxins produced by *Bt* allows the formulation of a variety of bioinsecticides by using the bacteria themselves or by expressing their toxin genes in transgenic plants. To date, many plant species have been genetically modified with *cry* genes, resulting in transgenic plants with a high level of resistance to insect [9]. However, it has been reported that several pests have developed resistance against Cry proteins [9, 10]. The current approach used to delay evolution of resistance to transgenic crops uses a "high dose" and "refuge" strategy [9, 11]. In addition, it is important to use a combination of *cry* genes and/or other genes encoding insecticidal proteins within the same transgenic crop [12, 13]. Due to extensive use of transgenic crops in developing countries based on *cry*-type genes, there is a need for alternative *cry* gene sequences to meet the challenge of novel insect resistance [7]. Crucial to this development is the identification of novel and more active strains with respect to insect pests of economically important crops [14].

The *cry* genes of *Bt* strains are known to be related to their toxicity [15, 16] and identification of these genes by means of PCR has been used to characterize and predict insecticidal activity of the strains [17, 18]. Nevertheless, a more complete characterization should include alternative methods. Phenotypical analysis such as protein profile determination provides useful information for typing and comparative studies [19]. The literature data report the possibility of using the whole-cell protein profile as a discriminating method with potency similar to RAPD with combined DNA patterns [1]. However, there is not always a good correlation between these factors and insecticidal activity of *Bt* strains [20, 21]. In addition, there is a need to develop knowledge about the biological properties and diversity of *Bt* isolates since these data allow a better understanding of the biological factors that determine insecticidal properties. Extracellular factors such as phospholipases, proteases and chitinases have shown to contribute to insecticidal activity of *Bt* [22].

During a screening programme of *Bt* isolates native to Argentina and toxic against Lepidoptera, several strains were characterized according to different biological parameters. In addition, promising isolates regarding their useful in biological control programmes -an environmentally safe technology of pest control- were exhaustively studied [14, 19, 23]. The present work showed most relevant results obtained during a course of those investigations. The discovery of highly pathogenic isolates against devastating insect pests reveals the usefulness of screening studies for novel *Bt* strains.

2. Biochemical characterization of *B. thuringiensis* isolates and assessment of toxicity

Crystalliferous spore-forming bacteria were isolated from both *S. frugiperda* larvae showing disease symptoms and soil samples collected in Argentina [19]. These samples came from maize, sorghum, wheat, grape or sugarcane cultivated fields. Briefly, larvae and soil sample suspensions were made in distilled water, heated at 80 °C for 15 min and then plated onto LB-agar. Plates were incubated at either 30 or 55 °C for 24 h. Colonies that did not grow at 55 °C were then analyzed for the presence of parasporal crystals by microscopic examination [24]. From a total of 254 colonies isolated from 490 different environmental samples, 14 were identified as crystal producer strains, giving a mean *Bt* index of 0.05. This result suggested that samples analyzed contained a high background level of other spore-forming bacteria. One crystalliferous strain came from sorghum cultivated field, while the others came from maize cultivated field. Concerning the source of isolation, 50% of crystal producers came from soil samples and the other 50% came from ill larvae. Interestingly, the last source provided the most pathogenic strains (Table 1).

Bacteria -characterized by conventional microbiological methods- possessed typical cellular and colonial morphologies, as well as physiological, biochemical and nutritional features that resembled *Bacillus* spp. They were motile and produced ellipsoidal endospores, located at sub-terminal position in the sporangia, and formed cream-colored colonies with irregular or circular edges on LB agar.

Phenotypical and molecular characterization	Strains														
	TRC11*	TMAN2*	THM8*	NN1**	TRC10*	RT**	TSA2*	TRC12*	N28**	MAN8**	MAN1**	THM30*	Bt 4D1***	LSM**	LQ**
Central spore	+	+	+	+	+	+	+	-	+	+	+	-	+	+	+
Sub terminal spore	-	-	-	-	-	-	-	+	-	-	-	+	-	-	-
Growth at pH 9	+	+	-	+	+	+	+	+	+	+	+	-	+	+	+
[b][e] Growth in 0.2 % chitin	-	+	+	+	-	+	+	+	-	+	-	-	+	+	+
[b][c][e] CMC hydrolysis (0.5 %)	+	+	+	+	+	-	-	+	+	+	-	+	+	-	-
[b][e] Chitin hydrolysis (0.2 %)	-	-	+	-	-	+	+	-	-	-	-	-	+	+	+
[b] Gelatin hydrolysis (12 %)	+	-	-	-	-	-	+	-	+	-	-	+	-	-	-
[b] Starch hydrolysis (2 %)	-	-	+	+	+	+	+	-	+	-	+	+	+	+	+
Gas production in glucose	-	+	-	+	-	+	-	+	-	-	-	-	+	+	+
[d] clindamycin	+	-	+	+	+	+	+	+	+	+	-	+	+	+	+
[d] gentamicin	+	+	-	-	+	-	-	+	+	-	-	-	+	-	-
[d] rifampicin	+	+	+	+	-	+	-	+	-	+	+	+	+	+	+
cry1	-	+	-	+	-	+	-	-	-	-	+	+	+	+	ND
cry2	-	+	+	-	+	+	-	+	+	-	-	+	+	+	ND
cry1Aa						-							+	+	
cry1Ab						+							+	+	
cry1Ac						+							+	+	
cry2Aa						-							+	+	
cry2Ab						+							+	+	

[a] Asterisks indicate the source of the isolates: *soil, **ill larvae and ***Bacillus Genetic Stock Center
[b] Expressed in w/v
[c] CMC: carboxy methyl cellulose
[d] Sensitivity to antibiotics was determined by using the routine diffusion plate technique. (+): sensitive and (-): resistant
[e] Growth on chitin and chitin hydrolysis were determined using colloidal chitin according to Kaur et al. (2005). This protocol was also used to analyze CMC hydrolysis
*Strains isolated from soil
**Strains isolated from ill larvae
***B. thuringiensis var. kurstaki 4D1 was provided by the Bacillus Genetic Stock Center (BGSC) (Columbus, Ohio) as well as the others Bt reference strains (see below).
ND: no determined (without amplification)

Table 1. Biochemical characteristics that presented variable response among the bacterial isolates. Molecular characterization is also showed (see below).

From a biochemical point of view, the 14 strains were catalase-positive, reduced nitrate and produced acetyl methyl carbinol in Voges-Proskauer broth; growth was observed at pH 7 on LB agar supplemented with 2, 3 and 5% NaCl and on LB agar at 30, 37 and 45 °C. The strains also hydrolyzed casein and were motile on soft LB agar. Negative results for all strains were obtained in several tests: no growth was observed on LB agar at pH 4 or at 50 °C and none of the strains hydrolyzed carboxymethyl cellulose (CMC) and urea. Antibiotic sensitivity tests revealed a resistance profile to penicillin, oxacillin, trimethoprim and a sensitive profile to erythromycin, vancomycin, levofloxacin, minocycline, chloramphenicol and teicoplanin. Phenotypic features that presented variability among the strains are showed in Table 1. The positive or negative result of each biochemical assay was entered in a 1-0 matrix. These data were subsequently analyzed through correspondence multivariate analysis, using Multivariate Statistical Package (MSVP) software (version 3.13). A cluster diagram based on these variable biochemical properties (that represented 54% of data variability) revealed that the strains formed two main groups (Figure 1). Group A comprised nine crystalliferous isolates which were clustered together with the reference strain *B. thuringiensis kurstaki* 4D1 (*Bt* 4D1). A second group (B) included three *Bt* strains while the remaining two strains presented more divergent features and hence were not included in any group. Isolates from the same sample and/or the same geographic region differed in their phenotypic features and consequently were not grouped together. This indicates that there was no clear association between *Bt* strains biochemical profile and the environments from which they were obtained [23].

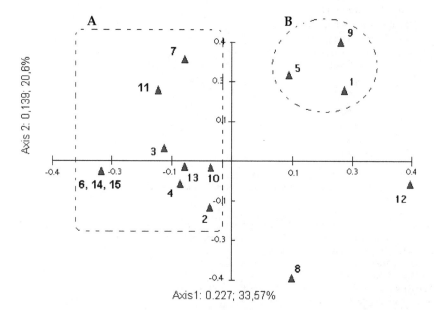

Figure 1. Correspondence multivariate analysis based on biochemical properties of *Bt* strains. 1: TRC11; 2: TMAN2; 3: THM8; 4: NN1; 5: TRC10; 6: *Bt* RT; 7: TSA2; 8: TRC12; 9: N28; 10: MAN8; 11: MAN1; 12: THM30; 13: **Bt* 4D1 (reference strain); 14: LQ and 15: LSM

Although most of the native isolates presented similar biochemical and phenotypical characteristics compared with reference strain *Bt* 4D1 (Group A) (Figure 1), they differed in their toxicity to *S. frugiperda*. Our results showed that the mortality on *S. frugiperda* neonate larvae was variable [19], ranging among values corresponding to Bt strains of bioinsecticides action low to moderate (Figure 2). However, strains named RT, LSM and LQ were found to be highly pathogenic, two of them, even more than reference strain *Bt* 4D1 which was selected for this analysis given it is the most widely used microorganism to control lepidopteran pests [25] (Table 2). This strong biological effect was represented by both a shorter LT_{50} and a higher mortality, which reached 100% in the case of RT strain on *S. frugiperda*, after five days of treatment. This result is extremely relevant considering that *S. frugiperda* is believed a pest with low sensitivity to *Bt* toxins [26]. In addition, when this strain was assayed against first instar larvae of *P. saucia*, reached 93% of mortality (Figure 3) suggesting that RT strain native to Argentina could possibly be employed in biological control of lepidopteran pests [19, 23]. It is important to stress that the high levels of mortality in the present work were obtained with a concentration of a spore-crystal suspension that was lower than some commercial *Bt* formulations; while our crystal spore suspensions presented a dose of 10^7 c.f.u. ml^{-1}, *Bt kurstaki* preparations generally present a dose of 10^9 c.f.u. ml^{-1} [27].

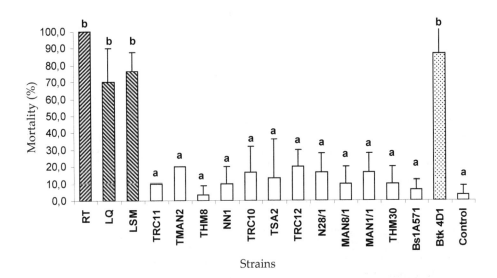

Figure 2. Insecticidal activity of crystalliferous native strains isolated from soil (☐) and ill larvae (◨).Bars sharing the same letter were not significantly different (*P* > 0.05, Tukey post-test). Reference strain: *Bt* 4D1 (⊡).Mortality was measured at the 7th day of assay. Ten individuals per treatment were observed and each treatment was repeated 10 times. *B. subtilis* 1A571 (*Bs* 1A571) was provided by the Bacillus Genetic Stock Center (BGSC).

Bt strain	*,a Mortality (%) ± SD	b LT50 (h) (95% fiducial limits)	*Specific biomass bound protease activity (± SD) (U g dry wt -1)
RT	100 ± 0 a	9.2 (10.4 –16.0)	1.98 ± 98 b
LSM	90.0 ± 7.3 a	37.7 (27.8 – 46.2)	1.80 ± 93 b
LQ	73.0 ± 5.7 c	79.6 (68.2 – 90.7)	1.14 ± 25 a
Bt 4D1	86.0 ± 15.1 b	58.7 (50.4 – 66.0)	946 ± 14 a
control	1.0 ± 3.1 d		

*Values followed by different letters were significantly different (P<0.05, Tukey post-test)
a Ten individuals per treatment were observed and each treatment was repeated ten times
b 50 % lethal time (LT50) was determined by Probit analysis. Mortality was scored every 24 h during seven days

Table 2. Comparison of mortality and 50% lethal time (LT50) of first instar larvae of *S. frugiperda* among native and reference *Bt* strains. Biomass-bound protease activity of pathogenic isolates are also showed.

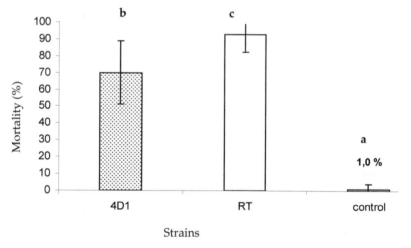

Figure 3. Comparison of insecticidal activity of *Bt* RT and the reference strain *Bt* 4D1 against the first instar larvae of *Peridroma saucia*. Mortality was measured at the 7th day of assay. Ten individuals per treatment were observed and each treatment was repeated 10 times. Mortality differed significantly between all treatments (*P* < 0.05, Tukey post-test).

3. Numerical analysis of insecticidal crystal proteins of *B. thuringiensis*

In order to differentiate native crystalliferous isolates and to evaluate the relationship between the toxicity assays against *S. frugiperda*, isolated bacteria and parasporal crystals were characterized by the whole-cell protein profiles in SDS-PAGE [19]. Protein bands were individually identified by their specific migration rates in the gels. Once bands were properly and distinctively identified, binary (0/1) matrices were constructed to compare the patterns. Electrophoretic analysis revealed the presence of 53 distinct bands with molecular weights ranging from 266 to 20 kDa (Figure 4). Numerical analysis clearly showed two distinct clusters (Figure 5). Cluster A comprised 11 isolations and the reference strain *B. thuringiensis* var. *thuringiensis* 4A4 (*Bt* 4A4) as well as those crystalliferous isolations that had no or very low toxicity against *S. frugiperda* first instar larvae. Interestingly, this group of native microorganisms produced proteins from 28 to 31 kDa but not proteins of ~135 and ~65 kDa. These lower molecular mass could correspond to Cyt toxins, entomocidal crystal proteins highly active against Diptera larvae [28]. On the other hand, all isolations with high toxic activity against *S. frugiperda* (RT, LSM and LQ strains) (Table 2) were located in cluster B, as well as the reference strains *Bt* 4D1 and *B. thuringiensis* var. *kurstaki* 4D3 (*Bt* 4D3). The isolation RT had a protein profile similar to *Bt* 4D1 with proteins of ~140 and ~70 kDa. Strains LSM and LQ showed protein bands of ~100 and ~81 kDa. These results demonstrate that the whole cell protein profiling not only allowed the differentiation of *Bt* at strain level but also revealed a possibility to apply protein profile analysis in classification of toxicity patterns.

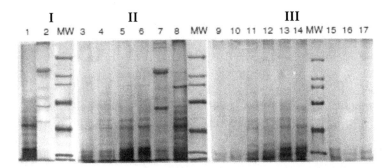

Figure 4. SDS-PAGE of whole-cell protein of crystalliferous strains. Gel I. Lines: 1: *Bt* 4D3, 2: *Bt* 4D1. Gel II. Lines: 3: N28, 4: *Bt* 4A4, 5: LSM, 6: LQ, 7: RT, 8: MAN8. Gel III. Lines: 9: THM8, 10: TMAN2, 11: THM30, 12: NN1, 13: TSA2, 14: MAN1, 15: TRC12, 16: TRC11, 17: TRC10. MW: Molecular weight marker Sigma-Aldrich were rabbit skeletal myosin (200 kDa), *E. coli* b-galactosidase (116.25 kDa), rabbit muscle phosphorylase b (97.4 kDa), Bovine serum albumin (66.2 kDa), hen egg white ovalbumin (45 kDa) and bovine carbonic anhydrase (31 kDa). Gels were stained with silver reagent. *B. thuringiensis* var. *kurstaki* 4D3 and *B. thuringiensis* var. *thuringiensis* 4A4 were also provided by the Bacillus Genetic Stock Center (BGSC).

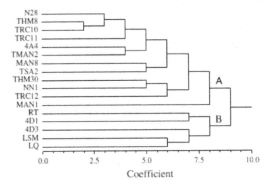

Figure 5. Dendrogram showing the relationship among *Bt* isolates based on electrophoretic whole cell protein patterns. Associations were produced using the simple matching coefficient and the neighbor-joining clustering method.

4. Molecular characterization of *B. thuringiensis* strains and crystal morphology

Although the presence of parasporal crystals is a diagnostic characteristic of *Bt* strains [1], the taxonomic identity of the toxic crystalliferous isolates was confirmed by amplification and partial sequencing of their 16S rDNA genes [19] (Table 3). The partial 16S rDNA sequences were tested by BLAST analysis against the GenBank data base. *Bt* LSM strain showed exact BLAST matches with the sequence from *Bacillus thuringiensis* var *kurstaki* (1 hit, 100% of identity, accession number EF638796). *Bt* LQ strain produced 4 hits (99% of identity, accession number EF638798), all of these corresponding to *Bt* species. Similarly, *Bt* RT (best hits, 13; 99% of identity, accession number EF638795) also shared a close relationship with others *Bt* strains, including *Bt* LSM.

Generally, *B. thuringiensis* insecticidal protein toxin genes (*cry*) reside on large self-transmissible plasmids, and individual *B. thuringiensis* strains can harbor a diverse range of plasmids that can vary in number from 1 to 17 and in size from 2 to 80 MDa [29,30], although it has also been suggested that they are present in the chromosome [31]. In this context, to study the plasmid profiles of *Bt* strains is an important parameter to determine their identity, since the number and size of these is associated with a particular *Bt* strain. Comparison between strains belonging to the same serotype showed a great difference in variability [30]. Some serotypes (e.g., israelensis) showed the same basic pattern among all its strains, while other serotypes (e.g., morrisoni) showed a great diversity of patterns. These results indicate that plasmid patterns are valuable tools to discriminate strains below the serotype level [30]. The profile of extrachromosomal elements in *Bt* is influenced by a number of stressful growth conditions, which determine its stability and heritability (e. i. high temperatures determine the plasmid loss), therefore it is neccesary to take some care . In this study, cultures were routinely grown at 30 °C to avoid this phenomenon. Detection and isolation of plasmid DNA was conducted following the method of Kado and Liu [32]. DNA plasmid samples were electrophoresed on 0.8 % (wt vol^{-1}) agarose gel. Our results showed that selected *Bt* strains present a complex plasmid profile (Figure 6).

In this experiment, the plasmid DNA was not linearized and therefore the same plasmid can produce as many as three different bands in the agarose gel. This made it difficult to determine the precise number of plasmids present in each complex plasmid profile. For this reason, we will refer to the number and size of plasmidic bands rather to plasmids themselves. An intense band above the chromosomal band (C) was observed in *Bt* RT, *Bt* LSM and *Btk* 4D1 suggesting that a large plasmid or plasmids is/are found in this strains which might be responsible for production of parasporal bodies. Compared with the other bacteria, *Bt* LQ presented a very different profile array, suggesting a different *cry*-genotype.

Identification of *cry* genes by means of PCR has been used to predict insecticidal activity of the strains [17,18] and to determine the distribution of *cry* genes within a collection of *B. thuringiensis* strains [20, 33]. In this context, our crystalliferous strains were characterized in terms of presence of *cry1* and *cry2* genes by amplification with general primers. The most toxic *Bt* strains RT, LSM and LQ were characterized through additional PCR with specific

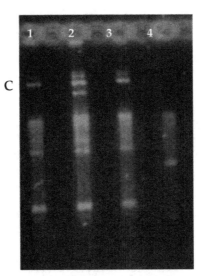

Figure 6. Plasmid profiles of *Bacillus thuringiensis* strains. Lanes: 1: *Bt* RT; 2: *Bt* 4D1; 3: *Bt* LSM; 4: *Bt* LQ. "C" indicate chromosomal DNA.

primers to identify the presence of *cry1Aa, cry1Ab, cry1Ac, cry1Ad, cry2Aa, cry2Ab* and *cry2Ac* genes (Table 3). PCR analysis showed presence of *cry1* and/or *cry2* genes in most of the isolates (Table 1). Specific PCR showed identical *cry* gene profile in both *Bt* LSM and the reference *Bt* 4D1, while *cry* gene content of *Bt* RT was different from them. DNA of *Bt* LQ was not amplified under the current reaction conditions (Table 1).

In addition, amplified fragments corresponding to *cry1* and *cry2* genes from *Bt* RT were sequenced and compared with *cry* genes sequences available from GenBank. This sequences had 99 and 95% identity with *cry1Ab* (EU220269) and *cry2Ab* (EU094885) genes, respectively. As shown in Figure 7, *cry1* and *cry2* partial sequences from *Bt* RT and *Bt* 4D1 were also aligned with five and six GenBank published *cry* sequences, respectively. The phylogenetic analysis revealed that *cry1* partial sequences from *Bt* RT and *Bt* 4D1 possess almost the same level of evolutionary distance (Figure 7A), while *cry2* partial sequence from *Bt* RT lies on a separate diverse branch not only of *cry2* from *Bt* 4D1 but also of the others analyzed *cry2* sequences (Figure 7B). Considering the phylogenetic analysis, it could be expected toxicity mediated by Cry1 rather than Cry2 crystal protein. In fact, *cry2* partial sequence from *Bt* RT shared a 95% homology with *cry2* sequence from a Colombian native *Bt* strain active against *Tecia solanivora* (Lep:Gelechiidae) (EU094885).

As mentioned before, *cry* genes are a family of genes associated with the toxicity of *Bt* against insects. While *cry1* encodes for proteins forming bipyramidal crystals and are related to toxicity to Lepidoptera [29] *cry2* encodes for cuboidal proteins, toxic to Lepidoptera and Diptera [39]. Our molecular and electron microscopy analyses of *Bt* RT are in agreement with all this evidence, since this highly pathogenic strain has both genes (Table 1) and both kinds of proteins (Figure 8A). In contrast, and although *Bt* LSM showed amplification products with *cry2* general and specific primers (Table 1) no cuboidal proteins were

Primer pairs	Nucleotide sequence	Reference
16S: 27F 1492R	5'-AGAGTTTGATCCTGGCTCAG-3' 5'-GGTTACCTTGTTACGACTT-3'	[34]
ITS: ISR-1494 ISR-35	5'-GTCGTAACAAGGTAGCCGTA -3' 5'-CAAGGCATCCACCGT-3'	[35]
Gral-*cry1*	5'-CTGGATTTACAGGTGGGGATAT-3' 5'-TGAGTCGCTTCGCATATTTGACT-3'	[36]
Gral-*cry2*	5'-GAGTTTAATCGACAAGTAGATAATTT-3' 5'-GGAAAAGAGAATATAAAAATGGCCAG-3'	[37]
Spe-*cry1*Aa	5'-TTATACTTGGTTCAGGCCC-3' 5'-TTGGAGCTCTCAAGGTGTAAA-3'	[38]
Spe-*cry1*Ab	5'-AACAACTATCTGTTCTTGAC-3' 5'-CTCTTATTATACTTACACTAC -3'	
Spe-*cry1*Ac	5'- GTTAGATTAAATAGTAGTGG-3' 5'- TGTAGCTGGTACTGTATTG-3'	
Spe-*cry1*Ad	5'-GTTGATACCCGAGGCACA-3' 5'-CCGCTTCCAATAACATCTTTT-3'	
Spe-*cry2*Aa1	5'-GTTATTCTTAATGCAGATGAATGGG-3' 5'-GAGATTAGTCGCCCCTATGAG-3'	[17]
Spe-*cry2*Ab2	5'-GTTATTCTTAATGCAGATGAATGGG-3' 5'-TGGCGTTAACAATGGGGGGAGAAAT-3'	
Spe-*cry2*Ac	5'-GTTATTCTTAATGCAGATGAATGGG-3' 5'-GCGTTGCTAATAGTCCCAACAACA-3'	

Table 3. Primer sequences used in this study.

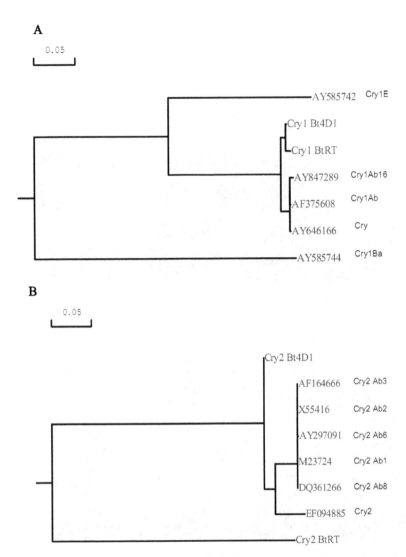

Figure 7. Phylogenetic rooted tree of *cry1* (A) and *cry2* (B) partial sequences from *B. thuringiensis* strains.

identified (Figure 8B). This suggests that a modification in the regulation of the gene would be responsible for the lack of protein product of this gene. Although the experimental growth conditions employed could also explain the lack of cuboidal proteins, the production of these proteins by *Bt* RT under identical experimental conditions argue against this possibility [14]. Cloning and sequencing the putative toxins with surrogate production made help clarify this issue as well as to confirm toxicity. In addition, *Bt* LQ showed no amplification products of *cry1* and *cry2* gene in several attempts (Table 1), despite the presence of bipyramidal crystals (Figure 8C). Noguera e Ibarra, [40] found that *cry* genes of a *Bt* strain isolated in Argentina that showed elongated bipyramidal crystals [41] presented 98% identity with *cry5Ba* genes. Therefore, *Bt* LQ may have Cry proteins other than Cry1 that form bipyramidal crystals.

From a methodological point of view, washing of crystal suspensions with absolute ethanol/distilled water (Figure 9B) was more appropriate for microscopic observation than washing with distilled water (Figure 9A).

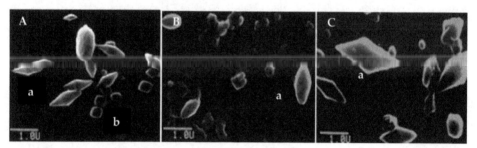

Figure 8. Scanning electron microscopy (SEM) of spore-crystal proteins from *Bt* strains. Concentrated spore-crystal suspensions were placed on a microscope lid and air-dried overnight. Samples were then coated with gold and examined using a scanning electron microscope. A) *Bt* RT; B) *Bt* LSM; C) *Bt* LQ. Both bipyramidal (a) and cuboidal (b) pesticidal crystal proteins are observed. Scale bar: 1 μm.

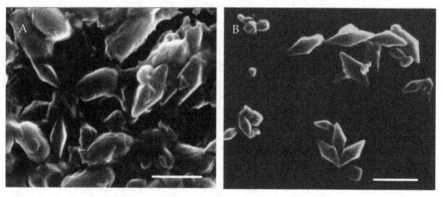

Figure 9. Scanning electron microscopy of spore-crystal proteins from *Bt* RT washed twice either with ethanol/water (1:1, v/v) (A) or with water (B). Scale bar: 1 μm.

As mentioned above, identification of *cry* genes by means of PCR has been used to predict insecticidal activity of the strains [17,18]. However, but since the primers are designed against known genes, the technique presents limitations in the search of novel *cry* genes. Moreover, the reliability of the prediction of insecticidal activity based on PCR results is dependent on the expression of the genes. In this context, a more complete characterization of *Bt* strains should include alternative PCR fingerprinting methods. Among them, assessment of length polymorphism of intergenic transcribed spacers (ITS) between the 16S and 23S rDNA genes has been shown to be an important tool for differentiating bacterial species and even prokaryotic strains [42]. In this context, ITS-PCR was performed as previously described by Daffonchio et al.[31]. Used primer pairs are showed in Table 3. Evaluation of ITS length polymorphism revealed an identical pattern among *Bt* RT, LSM and LQ strains and also with *Bt* 4D1 suggesting that ITS exhibited no polymorphism among the strains (Figure 10). In connection with this, Reyes-Ramirez and Ibarra [43] studied ITS profiles of 31 *Bt* strains and found them to be insufficient to discriminate between isolates.

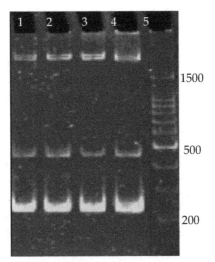

Figure 10. ITS-PCR of *B. thuringiensis* strains. Lanes 1: *Bt* LSM; 2: *Bt* LQ; 3: *Btk* 4D1; 4: *Bt* RT; 5: 100 bp DNA Ladder.

5. Assessment of enzyme activities in *B. thuringiensis*

Phenotypic characterization of selected strains allows identification of properties that are relevant at the moment of selecting bacteria for their use in environmental and agricultural microbiology. Synthesis of lytic enzymes by *Bacillus* species during the early sporulation phase is one of these properties. Secreted microbial enzymes may function as virulence factor that are essential for survival and spread in the host [44,45]. Our results indicate that the native *Bt* strains responded diversely regarding proteolytic, cellulolytic and/or chitinolytic activity (Table 1). Chitinolytic activity is a contributing factor in *Bt* pathogenicity [21] which our most pathogenic strains possessed. The enzymes involved would act on the peritrophic membrane of the host, which facilitates the entry of pathogens into the haemocoel of susceptible insects [21]. In addition, these strains showed no cellulolytic activity in medium supplemented with carboxymethyl cellulose (CMC), one of the products used as a matrix to protect *Bt* spores against high temperatures and UV exposure prevailing in natural environments [46]. This lack of cellulolytic activity is a desirable property given that gelled CMC will not be degraded at the time of *Bt* formulation, and therefore it can be employed for this purpose [13].

In *Bt*, high levels of protease activity are associated with both crystal and spore formation [47] and this activity may contribute in processing inactive Cry protoxin to active toxin [3].

While there is a reasonable understanding of soluble midgut proteases in toxin activation, little is known about the role of *Bt* protease in entomotoxicity. In this connection, during our investigations, biomass-bound protease and extracellular protease activities were determined in the toxic strains, which were process according to [48]. Proteolytic activity was assayed by using azocasein as substrate [49]. Table 2 shows that *Bt* RT, *Bt* LSM and *Bt* LQ displayed high biomass-bound protease activity. To our knowledge, the presence of this naturally immobilized enzyme activity has not been reported in *Bt* [18]. On the other hand, extracellular protease activity was observed when crude extracts of *Bt* strains were electrophoresed on SDS-PAGE containing gelatin powder [50]. The gels were then processed according to [51] for proteolysis to occur. Gel was stained with 0.1% (wt vol⁻¹) Coomasie Blue R-250. Proteinase K (10 mg ml⁻¹) was used for comparative analyze. All strains presented a clear zone of proteolytic activity which were larger in both *Bt* RT and *Bt* LSM (Figure 11). Although no correlation between protease activities and mortality values was initially detected, this result could be complementary information to consider in commercial *Bt* formulations, since the cell structure may act as a natural matrix able to protect the biomass-bound enzymes from the possible negative action of external agents; and therefore it could be that an increased percentage of *Bt* protease may actually reach the larvae midgut. Finally, it would be useful to explore the role of the extracellular and biomass-bound protease activities in crystal protein modification during *Bt* fermentation, the synergy of this protease source with insect entomotoxicity and the possible addition of vegetative cells in the final *Bt* formulation [18].

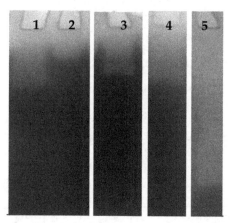

Figure 11. Identification of extracellular protease activity on 10% (wt vol⁻¹) SDS-PAGE containing 0.1% (wt vol⁻¹) gelatin powder. Lanes: 1: *Bt* RT; 2: *Bt* 4D1; 3: *Bt* LQ; 4: *Bt* LSM 5: Proteinase K.

Protein profiles are a useful tool to discriminate among strains, as they provide information about the proximity between species, subspecies and biovars [18, 52]. Considering this, characterization of microorganisms by means of their extracellular isoenzymes showing high polymorphism, as is the case of esterase, is particularly appealing. To determine extracellular esterase profiles, *Bt* strains were processed according to [44]. Briefly, strains were cultured on LB plates during 48 h at 30 ºC and crude extracts were recovered from solid media. Then, extracts were separated by native-PAGE. Esterase activity was assayed using 1.3 mM of α-naphthyl acetate (C2) derivative as substrate. Known electrophoretic esterase profiles of *Bacillus pumilus* A55 (EF638794.1) (*Bp* A55) were used for comparative analysis. The electrophoretic profiles of esterase activity showed differences among strains (Figure 12). *Bt* LQ showed a unique band/enzyme of 40 kDa as well as *Bt* LSM, but of about 60 kDa while *Bt* RT presented two bands of 95 and 60 kDa. Our results are in accordance with those by Norris [53], since it was possible to differentiate *Bt* strains by comparing the electrophoretic migration profiles of esterase produced during the vegetative growth phase (Figure 12) [13].

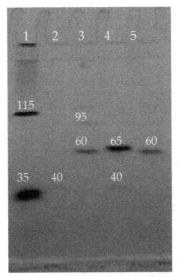

Figure 12. Enzymatic profile of esterase activity in native-PAGE 10%(wt vol^{-1}). Lanes: 1: *Bp* A55; 2: *Bt* LQ; 3: *Bt* RT; 4: *Bt* 4D1; 5: *Bt* LSM. Molecular weight of each band/enzyme is showed in kDa.

6. Conclusion

Lepidoptera causes some of the most devastating insect pests in important crops in America. Since economy of these regions depends largely of agriculture, their control is a priority as well as a necessity. In this context, use of environmentally safe technology to reduce crop damage like *B. thuringiensis* would be extremely valuable. Consequently, we set out to establish and characterize a collection of *Bt* isolates from samples collected in different Argentinean localities in order to find novel strains toxic against insect pests of economically important crops (like soybean and maize).

Fourteen *Bt* strains were isolated and phenotypically, genetically and biologically characterized. Analysis of larvicidal activity indicated that three strains exhibited high toxicity against lepidopteran larvae; this toxicity was in most, higher than that of the reference strain *Bt* 4D1.

The discovery of a highly toxic isolates reveals the usefulness of screening studies for novel *Bt* strains. The future application of these strains in biological control programmes requires optimization of the production conditions of the microorganisms using low-cost substrates. In this context, characterization of phenotypic and biochemical properties as evaluated in this study is highly relevant.

Author details

Analía Alvarez and Flavia del Valle Loto
Pilot Plant of Industrial and Microbiological Processes (PROIMI), CONICET, Tucumán, Argentina

Analía Alvarez
Natural Sciences College and Miguel Lillo Institute, National University of Tucumán, Tucumán, Argentina

7. References

[1] Konecka E, Kaznowski A, Ziemnicka J, Ziemnicki K (2007) Molecular and phenotypic characterization of *Bacillus thuringiensis* isolated during epizootics in *Cydia pomonella* L. J. Invertebr. Pathol. 94: 56-63.

[2] Bravo A, Gill S, Soberón M (2007) Mode of action of *Bacillus thuringiensis* Cry and Cyt toxins and their potential for insect control. Toxicon 49: 423-435.

[3] Brar S, Verma M, Tyagi R, Surampalli, R Bernabé S, Valéro J (2007). *Bacillus thuringiensis* proteases: Production and role in growth, sporulation and synergism. Process Biochem. 42: 773-790.

[4] Calderón, M, Alcocer González J, Molina M, Tames Guerra R, Rodríguez Padilla (2007). Adjuvant effects of crystal proteins from a Mexican strain of *Bacillus thuringiensis* on the mouse humoral response. Biologicals 35, 271-276.

[5] Virla E, Alvarez A, Loto F, Pera L, Baigorí M (2008) Fall Armyworm strains (Lepidoptera: Noctuidae) in Argentina, their associate host plants and response to different mortality factors in laboratory. Fla. Entomol. 91: 63-69.

[6] Moreno Fajardo O, Serna Cardona F (2006). Biología de *Peridroma saucia* (Lepidoptera: Noctuidae: Noctuinae) en flores cultivadas del híbrido comercial de *Alstroemeria* spp. Rev. Fac. Nal. Agr. Medellín. 2: 3435-3448.

[7] Berón C, Salerno G (2006) Characterization of *Bacillus thuringiensis* isolates from Argentina that are potentially useful in insect pest control. BioControl 51: 779-794.

[8] Gomes Monnerat R, Cardoso Batista A, Telles De Medeiros P, Soares Martins E, Melatti VM, Praca IB, Dumas VF, Morinaga C, Demo C, Menezes Gomes AC, Falcao R, Siqueira CB, Silva-Werneck JO, Berry C (2007) Screening of Brazilian *Bacillus thuringiensis* isolates active against *Spodoptera frugiperda*, *Plutella xylostella* and *Anticarsia gemmatalis*. Biol. Control 41: 291-295.

[9] Gassmann A, Carrière Y, Tabashnik E (2009) Fitness Costs of Insect Resistance to *Bacillus thuringiensis*. Annu. Rev. Entomol. 54: 147-63.

[10] Tabashnik B, Gassmann A, Crowder D, Carriére Y (2008) Insect resistance to *Bt* crops: evidence versus theory. Nat. Biotechnol. 26: 199-202.

[11] Tabashnik B, Carrière Y (2009) Environmental Impact of Genetically Modified Crops. In: Ferry N, Gatehouse A, editors. Insect Resistance to Genetically Modified Crops. Cambridge: CAB International.74-101pp.

[12] Christou P, Capell T, Kohli A, Gatehouse J, Gatehouse A (2006) Recent developments and future prospects in insect pest control in transgenic crops. Trends Plant. Sci. 11: 302-308.

[13] Sauka D, Benintende G (2008) *Bacillus thuringiensis*: generalidades. Un acercamiento a su empleo en el biocontrol de insectos lepidópteros que son plagas agrícolas. Rev. Argent. Microbiol. 40: 124-140.

[14] Alvarez A, Virla E, Pera L, Baigorí M (2011) Biological characterization of two *Bacillus thuringiensis* strains toxic against *Spodoptera frugiperda*. World J. Microbiol. Biotechnol. 27: 2343-2349.

[15] Carozzi N, Kramer V, Warren G, Evola S, Koziel MG (1991) Prediction of insecticidal activity of *Bacillus thuringiensis* strains by polymerase chain reaction product profiles. Appl. Environ. Microbiol. 57: 3057-3061.

[16] Padidam M (1992) The insecticidal crystal protein Cry1A (c) from *Bacillus thuringiensis* is highly toxic for *Heliothis Armigera*. J. Invertebr. Pathol. 59: 109-111.

[17] Ben-Dov E, Zaritski A, Dahan E, Barak Z, Sinai R, Manasherob R, Khamraeb A, Troitskaya E, Dubitsky A, Berezina N, Margalith Y (1997) Extended screening by PCR for seven *cry*-group genes from field-collected strains of *Bacillus thuringiensis*. Appl. Environ. Microbiol. 63: 4883-4890.

[18] Hansen B, Damgaard P, Eilenberg J, Pedersen JC (1998) Molecular and phenotypic characterization of *Bacillus thuringiensis* isolated from leaves and insects. J. Invertebr. Pathol. 71: 106-114.

[19] Alvarez A, Virla E, Pera L, Baigorí M (2009a). Characterization of native *Bacillus thuringiensis* strains and selection of an isolate active against *Spodoptera frugiperda* and *Peridroma saucia*. Biotechnol. Lett. 31: 1899-1903.

[20] Porcar M, Juarez-Perez V (2003) PCR-based identification of *Bacillus thuringiensis* pesticidal crystal genes. FEMS Microbiol. Rev. 26: 419-432.

[21] Martínez C, Ibarra J, Caballero P (2005) Association analysis between serotype, *cry* gene content, and toxicity to *Helicoverpa armigera* larvae among *Bacillus thuringiensis* isolates native to Spain. J. Invertebr. Pathol. 90: 91-97.

[22] Soberón M, Bravo A (2001) *Bacillus thuringiensis* y sus toxinas insecticidas. In: Microbios en línea. UNAM México. Available vía DIALOG http://www.biblioweb.dgsca.unam.mx/libros/microbios /Cap12 Accessed 1 March 2011.

[23] Alvarez A, Pera L, Loto F, Virla E, Baigori M (2009b) Insecticidal crystal proteins from native *Bacillus thuringiensis*: Numerical analysis and biological activity against *Spodoptera frugiperda*. Biotechnol. Lett. 31: 77-82.

[24] Sharif F, Alaeddinoğlu N (1988) A rapid and simple method for staining of the crystal protein of *Bacillus thuringiensis*. J. Ind. Microbiol. 3: 227-229.

[25] Arango J, Romero M, Orduz S (2002) Diversity of *Bacillus thuringiensis* strains from Colombia with insecticidal activity against *Spodoptera frugiperda* (Lepidoptera: Noctuidae). J. Appl. Microbiol. 92: 466-474.

[26] Del Rincón-Castro M, Méndez-Lozano J, Ibarra J (2006) Caracterización de cepas nativas de *Bacillus thuringiensis* con actividad insecticida hacia el gusano cogollero del maíz *Spodoptera frugiperda* (Lepidoptera: Noctuidae). Folia Entomol. Mex, 45: 157-164

[27] Seligy V, Rancourt J (1999) Antibiotic MIC/MBC analysis of *Bacillus*-based commercial insecticides: use of bioreduction and DNA-based assays. J. Ind. Microbiol. Biotechnol. 22: 565-574.

[28] Gough J, Kemp D, Akhurst R, Pearson R, Kongsuwan K (2005) Identification and characterization of proteins from *Bacillus thuringiensis* with high toxic activity against the sheep blowfly, *Lucilia cuprina*. J. Invertebr. Pathol. 90: 39-46.

[29] González J, Carlton B. (1980) Patterns of plasmid DNA in crystalliferous and acrystalliferous strains of *Bacillus thuringiensis*. Plasmid 3:92–98.

[30] Reyes-Ramírez IJ (2008) Plasmid patterns of *Bacillus thuringiensis* type strains. Appl. Environ. Microbiol. 74: 125-129.

[31] Kronstad J, Schnepf H, Whiteley H (1983) Diversity of locations for *Bacillus thuringiensis* crystal protein genes. J. Bacteriol. 154:419–428.

[32] Kado C, Liu S (1981) Rapid procedure for detection and isolation of large and small plasmids. J. Bacteriol. 145:1365-73.

[33] Chak K, Chao D, Tseng M, Kao S, Tuan, S Feng, T (1994) Determination and distribution of *cry*-type genes of *Bacillus thuringiensis* isolated from Taiwan. Appl. Environ. Microbiol. 60:2415-2420.

[34] Weisburg W, Barns S, Pelletier D, Lane D (1991) 16S Ribosomal DNA amplification for phylogenetic study. J. Bacteriol. 173, 697-703.

[35] Daffonchio D, Borin S, Frova G et al (1998) PCR fingerprinting of whole genomes: the spacers between the 16S and 23S rRNA genes and of intergenic tRNA gene regions reveal a different intraspecific genomic variability of *Bacillus cereus* and *Bacillus licheniformis*. Int. J Syst Bacteriol 48:107–116.

[36] Bravo A, Sarabia S, Lopez L, Ontiveros H, Abarca C, Ortiz A, Ortiz M, Lina L, Villalobos FJ, Peña G, Nuñez-Valdez ME, Soberón M, Quintero R (1998) Characterization of *cry* genes in Mexican *B. thuringiensis* strain collection. Appl. Environ. Microbiol. 64: 4965-4972.

[37] Ibarra J, del Rincón M, Ordúz S, Noriega D, Benintende G, Monnerat R, Regis L, Oliveira C, Lanz H, Rodriguez M, Sánchez J, Peña G, Bravo A (2003) Diversity of

Bacillus thuringiensis strains from Latin America with insecticidal activity against different mosquito species Appl. Environ. Microbiol. 69:5269–5274.

[38] Cerón J, Covarrubias L, Quintero R, Ortiz M, Aranda E, Lina L, Bravo A (1994) PCR Analysis of the *cryI* Insecticidal Crystal Family Genes from *Bacillus thuringiensis*. Appl. Environm. Microbiol. 60:353-356.

[39] Al-Momani F, Saadoun I, Obeidat M (2002) Molecular characterization of local *Bacillus thuringiensis* strains recovered from Northern Jordan. J. Basic Microbiol. 2: 156-161.

[40] Noguera P, Ibarra J (2010) Detection of New *cry* Genes of *Bacillus thuringiensis* by Use of a Novel PCR Primer System. Appl. Environ. Microbiol. 76: 6150-6155.

[41] Benintende G, López-Meza J, Cozzi J, Ibarra JE (1999) Novel non-toxic isolates of *Bacillus thuringiensis*. Lett .Appl. Microbiol. 29: 151-155.

[42] Daffonchio D, Raddadi N, Merabishvili M (2006) Strategy for identification of *Bacillus* cereus and *Bacillus thuringiensis* strains closely related to *Bacillus anthracis*. Appl. Environ. Microbiol. 72:1295–1301.

[43] Ramírez R, Ibarra J (2005) Fingerprinting of *Bacillus thuringiensis* type strains and isolates by using *Bacillus cereus* group-specific repetitive extragenic palindromic sequence-based PCR analysis. Appl. Environ. Microbiol. 71: 1346-1355.

[44] Stefanova M, Leiva M, Larrinaga L, Coronado M (1999) Metabolic activity of *Trichoderma* spp. isolates for a control of soilborne phytopatogenic fungi. Rev. Fac. Agron. (LUZ) 16: 509-516.

[45] Lengyel MJ, Pekár S, Felföldi G, Patthy A, Gráf L, Fodor A, Venekei I (2004) Comparison of Proteolytic Activities Produced by Entomopathogenic *Photorhabdus* Bacteria: Strain- and Phase-Dependent Heterogeneity in Composition and Activity of Four Enzymes. Appl. Environ. Microbiol. 70: 7311-7320.

[46] Cokmus C, Elcin M (1995) Stability and controlled release properties of carboxymethylcellulose-encapsulated *Bacillus thuringiensis* var. israelensis. Pest. Sci. 45: 351-355.

[47] Ibarra JE, Del Rincón Castro MC, Galindo E, Patiño M, Serrano L, García R, Carrillo JA, Pereyra-Alférez B, Alcázar-Pizaña A, Luna-Olvera H, Galán-Wong L, Pardo L, Muñoz-Garay C, Gómez I, Soberón M, Bravo A (2006) Los microorganismos en el control biológico de insectos y fitopatógenos. Rev. Latinoam. de Microbiol. 48: 113-120.

[48] González C, Martínez A, Vázques F, Baigorí M, Figueroa L (1996) New method of screening and differentiation of exozymes from industrial strains. Biotechnol. Tech. 10: 519-522.

[49] Secades P, Guijarro A (1999) Purification and Characterization of an Extracellular Protease from the Fish Pathogen *Yersinia ruckeri* and Effect of Culture Conditions on Production. Appl. Environ. Microbiol. 65: 3969-3975.

[50] Heussen C, Dowdell E (1980) Electrophoretic analysis of plasminogen activators in polyacrilamide gels containing sodium dodecyl sulfate and copolymerized substrates. Anal. Biochem. 102: 196-202.

[51] Ferrero M, Castro G, Abate M, Baigorí M, Siñeriz F (1996) Thermostable alkaline proteases of *Bacillus licheniformis* MIR 29: Isolation, Production and Characterization. Appl. Microbiol. Biotechnol. 45: 327-332.

[52] Berber I (2004) Characterization of *Bacillus* species by numerical analysis of their SDS-PAGE protein profiles. J. Cell Mol. Biol. 3: 33-37.

[53] Norris J (1964) The classification of *Bacillus thuringiensis*. J. Appl. Bacteriol. 27:39- 447.

Chapter 6

Tropical Forest and Carbon Stock's Valuation: A Monitoring Policy

Thiago Metzker, Tereza C. Spósito, Britaldo S. Filho,
Jorge A. Ahumada and Queila S. Garcia

Additional information is available at the end of the chapter

1. Introduction

Carbon is the fourth most abundant element on Earth. It is estimated that the world's forests store 283 gigatonnes (1Gt = 1 billion tons) of carbon in their biomass alone and 638 Gt of carbon in the ecosystem as a whole (to a soil depth of 30 cm). Thus, forests contain more carbon than the entire atmosphere. Carbon is found in forest biomass and dead wood, as well as in soil and litterfall [1]. Consequently, changes in forest carbon storage, resulting from a shift in land use, have a significant impact on global climate change [2].

Changes in climate occur naturally, through processes operating on a geologic time scale. For example, the main species presently inhabiting the planet have survived climate changes during the Pleistocene, adjusting their geographical distribution to weather conditions. However, the speed and magnitude of changes that have been occurring in the Earth's climate system since the Industrial Revolution are currently of great concern. In 1991, the Intergovernmental Panel on Climate Change (IPCC) published a first report about global temperature increases caused by the intensification of the greenhouse effect. After this official announcement, the IPCC has established different working groups with scientists from various parts of the world in order for them to meet and compile as much information as possible and to update scientific predictions about the climatic future of the planet. The reports that have been produced by the international scientific community are considered as the main reference for global climate change.

Currently, scientific societies question the capacity of the present biota to tolerate such changes, in an environment that has been highly fragmented by human intervention and where what is still left intact is confined within protected areas. Changes within biota can result in changes in the ecosystem services they provide. Human well-being depends directly and indirectly on the environmental services provided for free by the natural world, including climate regulation, soil formation, erosion control, carbon storage, nutrient cycling, provision of water (both quality and quantity), maintenance of hydrological cycles, preservation of genetic resources, scenic beauty, among others [3]. Furthermore, tropical forests contain 50% of all world species and are considered mega-diverse environments. Therefore, changes in any of these services can have serious consequences for biodiversity, for the natural carbon cycle and the hydrological cycle, which may in turn alter the world economy and affect the everyday life of humans and other species on the planet.

How can these changes be monitored? One way to monitor biodiversity and carbon stocks over large areas is through the establishment of forest inventories. These are effective tools for estimating the type, amount and condition of forest resources over large areas [4]. The regular collection of measurements within Permanent Monitoring Plots (PMPs), combined with the use of statistical techniques, provide a baseline for assessing changes in the structure and dynamics of a forest and permit the construction of predictive models [5]. In the last decade, there has been a large increase in the installation of PMPs in different tropical forest sites around the world, especially in the Amazon Rainforest, where large monitoring networks (TEAM, PELD, CTFS, RAINFOR, LBA, REDEFLOR, PDBFFE and CIFOR) have been established. These programs increase the level of understanding of ecological systems, transforming the knowledge base [6]. However, there are still serious deficiencies in estimating carbon stocks and other components of other types of tropical forests, types and others components of tropical forests.

Within the current political and environmental international situation it is vital that all countries, whether or not signatories of the Kyoto Protocol, do promote initiatives to monitor their biodiversity and their carbon stocks. These data are strategic for each country because they indicate where and how the management of natural resources can bring benefits to local people (local scale), they support the creation of public policies that can become part of the country's legislation (regional scale) and promote policies for adaptation to an increased vulnerability to climate change (global scale).

This chapter, "Tropical Forest and Carbon Stock's 1 Valuation: A Monitoring Policy", incorporates parts of the TEAM (Tropical Ecology Assessment Monitoring) protocol [7] and the knowledge generated over six years of monitoring permanent plots in an area of the Atlantic Rainforest in Brazil. It aims to discuss the importance of planning and implementation of PMPs, the main techniques used, and the errors associated with them. Biomass, carbon stock calculation techniques and data analysis will also be discussed, among other topics. Data collection and analysis have a greater value when incorporated into natural resource management policies, such as Payment for Environmental Services (PES), which are provided by nature. A comprehensive approach involving stakeholders at all levels, from the local to the global scale, is essential for the success of integrated policies. Each of the topics listed below will be presented with the aid of practical examples, figures and tables, in order to allow readers the opportunity to fully engage with the subject matter and, most importantly, to begin to understand how to apply these practices in their own social and environmental contexts.

2. Inventory Data

2.1. Methods for establishing Permanent Monitoring Plots (PMPs)

The establishment of vegetation monitoring networks is a strategy that aims to develop an integrated database through systematized collections using a single monitoring protocol on various sites. In the vegetation network implementation, it is extremely important that the database management team be clear about the questions to be asked and the objectives for the collection of field data. This systemization has implications directly related to the method of collection and the definition of the protocol for implementation and monitoring. The primary analyses to be conducted also must be predefined as they too have a direct impact on the sample design and the means of data collection.

During the planning of a monitoring network, it is important to keep in mind that the key objective is to conduct large-scale analyses that can speak to physiognomy, biomes and wider generalizations. This scale of work is fundamental in order to accomplish robust analyses and to study broad-scale ecological processes. However, it should be noted that local and regional data and publications are also part of this network as they promote the development of local scientific knowledge, along with the participation of the team responsible for the collection of field data. These initiatives encourage cooperation and sharing of experience, in addition to motivating those who are responsible at the local level to continue the work of monitoring once the objectives and results of the initiative are made clear to all involved.

The means of disseminating results should also be defined in the planning phase. For example, during this phase, contact can be made with the editors of scientific journal where there is an intention to publish, in order to establish a connection with the journal and develop credibility for a strong relationship. The sharing of the monitoring protocol, the initial results and the key conclusions at national and international conferences provides visibility for the project and stimulates ongoing discussions with other researchers in the topic area. This interaction and sharing of experience always benefit the project as they increase quality and strengthen key elements. The network planning team should also identify other forms of communication for scientific dissemination, such as specialized documentaries, news networks, community sites and scientific blogs. These promote dissemination and constructive discussion of the conclusions and methods of the published initiative. Another tactic that can make a significant contribution to successful monitoring over the long term by strengthening relationships with local teams is the development of news releases in the local language where the data was collected.

As with any good plan, the protocol must be rigorous. Several protocols for monitoring tropical forests are available including RAINFOR's [8], TEAM's [7] and the Smithsonian's Center for Tropical Forest Science [9]. However, it must also be flexible enough to be adapted and to evolve naturally according to the knowledge generated during the planning process, as well as to the local reality of each site. Ongoing workshops with the local team guarantee that acquired experience is formally recorded, in addition to facilitating the continuous improvement of the protocol by applying experience acquired through its execution *in situ*.

2.2. Geoprocessing techniques for area selection

Many field procedures involve high costs due to transportation and logistics. Therefore, prior to any field procedure, errors in area selection can be minimized by careful planning using GIS techniques. In addition to playing an important role in the preliminary phase (planning), these tools are also very useful in the data analysis phase. When these instruments are used extensively by a qualified professional, significant economies of time and financial resources can be achieved.

After clearly defining the objectives for the implementation of the monitoring network, the next phase is the selection of potential areas to house the plots. The use of GIS allows for a more confident selection of the target areas since it works with georeferenced bases and shapes which allow for simulation of PMPs implemented in practically any location in the world. These areas can be selected by process of elimination from those that, for example, do not have the required attributes or by selection of multiple criteria that involves interpolation of various bases. Through experience acquired in the implementation and monitoring of PMPs, we understand that the minimal criteria for exclusion of target areas for monitoring include:

- Areas that possess accentuated declivity;
- Areas that are not easily accessible and complicate field logistics;
- Areas with creeks, swamps, lakes and rivers;
- Areas that have significant spatial heterogeneity;
- Areas that have variations in the type of soil.

Assuming that the objective of monitoring is to evaluate the temporal dynamics of primary vegetation areas, the areas that are not located in Conservation Units can be excluded first. It is understood that forested areas protected by law in any part of the world represent the highest percentage of protected primary areas. After this first filter, the layers or shapes that meet the exclusion criteria cited above are applied. This type of cut is made relatively quickly, while still in the office, but can reduce a universe of potential samples by more than 90% in certain regions of the world, thus optimizing the accuracy and use of the project's financial resources.

Following elimination of the areas not selected for the sample, the professional responsible for the GIS technology should create polygons capable of housing the future PMPs so that random samples can be selected from within the universe of possible options, thus establishing statistical confidence for the sample. Another important point is that the PMPs should be replicated in areas where there is similar physiognomy, so that means, errors and reliable statistics can be obtained.

It is of fundamental importance for the field team that thematic maps be developed by the GIS team. These maps should be easy to visualize and understand, with current satellite images and superimposed colored sketches of the PMPs in various layers. Essential factors for successful field work include the standardization of symbols, language and scale of work, as well as pre-definition of a standard datum, and being in a system of unique coordinates compatible with the use of local GPSs. The field maps should also be plasticized to avoid stains and tears which can often occur with the use of these materials in the middle of the forest.

2.3. Choosing target areas

The field team should also be very clear about the objective of monitoring. When the project's primary issue is related to the dynamics of areas in recovery or to the differences between primary and secondary vegetation areas, area selection involves different parameters. When the question is focused on temporal variations in areas of intact vegetation in the climactic stage, area selection will be directed primarily toward areas protected by legal mechanisms in each region, ensuring that there will be no interference in the plot throughout the years of monitoring. Depending on the objective, criteria for inter-site analysis can also be established, such as a latitudinal gradient temperature or rainfall gradient, soil gradient, etc.

Once all of the criteria have been established, the field team should depart in order to locate and validate the target areas *in situ*. In addition to being accompanied by local guides, the team should be supplied with basic field supplies as well as thematic maps developed by the GIS team, a GPS, a compass, and a camera for the validation or invalidation of areas previously defined by the GIS team. Additionally, the field team should have in your GPS all points and layers that were previously prepared by the GIS team. For example, see [10] for a complete data transfer protocol.

It is important that the field team be fully trained on the monitoring protocol and have the ability to independently decide at any given moment if an area truly possesses the defined selection criteria or if it would be better to search for a new area. This decision is a key since all monitoring throughout the years ahead will depend on the correct choice and demarcation of these plots. In order to select the best areas for PMPs to be implemented, various factors should be taken into consideration, including the homogeneity of the forest typology to be sampled, the existence of water courses, logistics, access, type of soil and inclination of the terrain.

Due to difficulties of orientation and localization in interior bush areas, the geographical coordinates should be checked and the location of the field team confirmed upon arrival at the target area. Once the location has been verified, a marker should be placed in the ground (a PVC tube of about 1.3 m can be used) to be the point of coordinates 0,0 (X, Y), which will serve as a reference point for the validation of the area as well as for future plot implementation. This point will be used to evaluate the area to decide whether or not it will be selected for PMP implementation. Thus, using a compass, the direction of the course should be read, so that the angle of the directions has a difference of 90° (straight angle). The course is followed in the first direction (X), remaining aligned with the lead angle on the compass, stopping every 20 meters to check the coordinates and the direction of the course. In the field, detours are very common during a walk/hike due to natural obstacles such as fallen trees and branches, the presence of lianas or holes in the ground, or large trees that have to be circumvented. It is important in this verification phase, as well as in the PMP implementation phase, that knives and scythes are not to be used to open trails or forest passages as they can have a long term impact with significant implications on the dynamics of vegetation. Thus, when faced with a natural obstacle, the ideal would be for the team to circumvent it and return to the defined course in order to continue with area verification.

The team should be aware of sudden changes in the type of soil, the existence of accentuated declivity that was not possible to identify in the satellite images, or any other element that strongly differentiates the landscape and that could negatively impact the monitoring or the homogeneity of the plot. This should be recorded in a designated worksheet in order to justify the decision not to use the area in question. Once line X has been verified, the same procedure is conducted with line Y beginning from ground zero. If an area does not possess significant heterogeneity, the selection of the plot must be validated, assigning a number and a syllable to be used throughout the entire period of monitoring and analysis of that area (e.g. 01-LP).

2.4. Implementation of PMPs in the field

Once the entire validation process is complete, the actual marking of the PMP in the field is undertaken. On the day prior to departure, a checklist should be reviewed of all equipment required for field implementation, such as PVC tubes, rubber hammer, colored tape, polypropylene cord, compass, GPS, binoculars, clipboard, collection worksheets, plastic bags, masking tape, pencils, erasers and pens. In addition to support materials, specialized clothing must also be taken, such as boots, leggings and field jackets (with many pockets). The PMP implementation team should be comprised of at least 4 people, primarily to divide the weight of materials to be taken to the selected PMP area, as the tubes or stakes used to mark the chosen spots are very heavy and bulky.

Upon arrival at the PMP location previously marked as 0,0, a suitable location to leave all of the equipment should be identified, as well as an appropriate place to have snacks or lunch while in the field. This location, named "Support Station - SS" should be located in the outlying area of the PMP so that it does not interfere with the vegetation to be monitored on the plot. The ground should be covered by a light blue tarp (or any color that strongly contrasts the forest floor), upon which all of the equipment should be placed to avoid loss. Again, it is imperative that the team be careful not to allow any type of vegetation (lianas, branches or shrubs) to be cut during plot implementation.

In the following example, we simulate the implementation of a 1 ha PMP (10.000 m²) according to the TEAM protocol for vegetation monitoring [7]. The size of the PMP will depend on the initial objective outlined by the team responsible for managing the project. The size of 1 ha is widely used in permanent plots whose objectives are related to monitoring the dynamics and carbon stocks for the site in question.

Starting at 0,0, two baselines (X, Y) should be projected, at 90° perpendicular angles, which will serve as reference points throughout the PMP implementation. Each baseline should be spiked every 20 meters, with their distance verified using a measuring tape and direction verified by reading the course angle on the compass. After the 6 spikes for each baseline have been duly marked and inserted into the ground, the entire line should be measured to confirm its length, which should be a total of 100 meters. Each spike placed every 20 meters should be sequentially numbered, as well as having its Cartesian coordinates on the plot recorded (e.g. 20, 0; 40, 0; 60, 0;…). Once line X has been completed, the formation of line Y can be undertaken using the same procedures previously followed.

Once the two baselines have been formed, the internal squares of the PMP can be developed. In order to close a PMP, two basic methods can be used: creating 5 lines parallel to baseline Y (Figure 1-B) or creating small 400 m² squares, forming sequential lines until the entire PMP is closed (Figure 1-A).

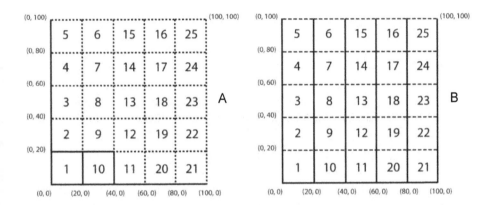

Figure 1. Sample structure of the Permanent Monitoring Plot (PMP), with 25 sub-plots. A – Means of closure using squares; B – Means of closure using lines. Adapted from TEAM (2010).

2.5. Marking trees

After marking the PMP, the individuals to be monitored are marked and data collection is undertaken. For studies related to long-term monitoring of the structure and dynamics of vegetation, it is common for the sample to include all individuals in the forest that have DBH ≥ 10 cm (Diameter at Breast Height). For studies of biomass and carbon stocks, individuals with DBH < 10 cm are not included, due to their low contribution to the total stocks of the PMP. In general, if the objective is to monitor changes in floristic composition and the biodiversity of the PMP, these smaller individuals should be incorporated into the monitoring.

In this case, all of the trees palms and lianas with a DBH greater than or equal to 10 cm should be marked and measured. The POM (Point of Measurement) is the point on the tree or liana where their respective diameters are measured. The POM is marked at 1.30 m with the help of a PVC tube graded at 1.60 m and 1.30 m to avoid error related to the different heights of the field markers. However, for individuals with tabular roots, sapopemas or buttress roots, the POM should be identified at 50 cm above the highest root (Figure 2). This is a valid change since it is common in forest inventories to find all stems with their DBHs measured at 1.30 m. When these data are inserted into allometric equations to calculate biomass, they overestimate biomass, increasing the standard error of these calculations [9-13].

In the case of trees that have many deformities at the POM, a modular ladder up to 12 meters (4 modules of 3 meters each to make it easy to transport in the forest) should be used so that the best location on the tree can be selected for diameter measurement (Figure 2 and Figure 3). Leaning or fallen trees should have their DBH measured following the methodology above; however, the distance from the base should be measured from the underside of the tree (Figure 2) in order to obtain an accurate distance. For trees with multiple trunks, where forking occurs below 1.30 m, each trunk should be considered a separate individual (Figure 2), with the number of measurements matching the number of trunks for the tree.

Once the best area for DBH measurement has been selected, it should be painted with yellow paint. This can also be done with a type of stamp (stencil) that can be made out of a sheet of hard plastic that is cut in the center in the following dimension: 20 x 3 cm. After selecting the location to be painted, the stamp (stencil) is placed on the tree and the POM is painted (Figure 3). In addition to facilitating field work, this stamp also standardizes the width of the paint marking on the trees, thus reducing the possibility of errors in future tree measurements.

This marking should be re-done every two years so that the specific POM is not lost. In order to avoid errors related to POM marking, the height at which the POM is marked should be recorded in a designated field worksheet. This procedure, along with painting the POM, guarantees that the measurement will be done at the exact same point during re-census throughout the monitoring period.

All of the individuals selected should be marked with nails and aluminium tags using increasing numbers according to the layout within the PMP. The nail should always be a distance of 40 cm from the POM so that the nail hole does not damage the trunk and consequently alter the POM. It is very common to see trees in the forest that have significant deformities resulting from a small nail hole. Bacteria and pathogens can enter through this small orifice and cause significant stress to tree trunks. Another important point is that the nail should be pointed downward whereby the tag is touching the head of the nail, since it is common to see trees that envelop around the tags over time when the tags had been touching the trees themselves.

After numeration and marking are complete, each individual should be identified at the highest taxonomic level possible in the field. It is highly recommended that photos be taken of the collected branches and that a collection of each species within the PMP be maintained as a botanical collection specific to each region. The data should be recorded in field worksheets and branch samples that are not identified should be taken for laboratory activities, herbarium consultations and completion of taxonomic identification by specialists. All field collections should be labelled with masking tape, recording their PMP number and reference code.

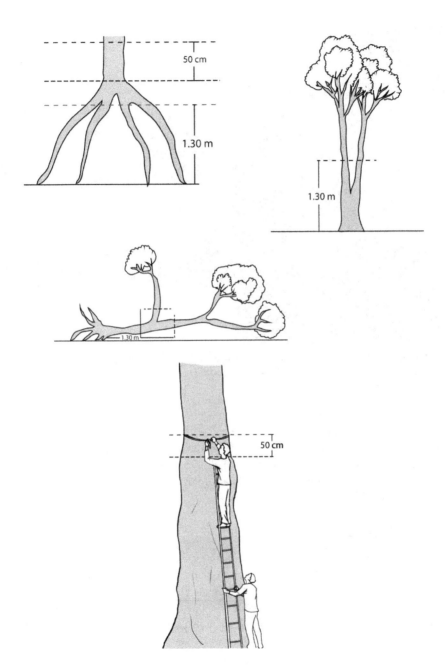

Figure 2. Details for marking trees with deformities in the field. A – For tabular roots, the POM is measured 50 cm above the last root; B – For multiple trunks, each is measured as a separate individual, provided the forking is below 1.30 m; C – For fallen trees, the distance is taken from the underside; D – For tall trees, the measurement should be done with the support of modular ladders.

Figure 3. Details of marking big trees at Rio Doce State Park: Use of a modular ladder up to 12 meters and POM painting process using a stencil. Source: Metzker, T.

With the collection and identification of botanical material, local guidebooks can be developed for the identification of trees registered within the PMPs. The guidebook could include photos of dried plants, taxonomic identification, location of the species, whether or not there are medicinal purposes, and details about flowers or fruits. In collaboration with local experts, the production of this type of material strengthens relationships between project managers and the execution team, in addition to producing registered material that is easily understood by the local population.

2.6. Calibration of diameter tape

As a result of the measurement process, diameter tape can become stretched or it may come from the factory already with small defects. Considering that the annual growth rate of a tree stratum in the forest is ~0.2 cm/year [14] small measurement errors can have a strong impact on the final results. In order to avoid this type of error, the diametric tape should be calibrated using an aluminium ruler prior to each census, thus maximizing the level of precision in the results.

2.7. Measurement calibration

Errors in reading the diametric tape or errors in the position of the tape on the tree can be common during the census, negatively impacting the processing and analysis of data.

Therefore, prior to each census, it is also necessary to calibrate the technician responsible for measuring the trees.

On the first day of the census, all possible measurements should be completed within a given PMP. One or two days later, the same person who measured the trees on day one should return to the same area and re-measure all of the previously measured trees. The results are considered good if the one measuring the trees obtains a minimum of 70% accuracy, or 90% with less than 1 mm of error. If these parameters are not reached, the procedure is repeated, even with others doing the measuring, until the required precision is obtained. The objective in each phase is to minimize potential errors that generally occur in field activities and which substantially impact data analysis.

2.8. Census and re-census

The measurement of the individuals located in a PMP is the heart of the entire initiative. The measurements conducted during the first census should be done with careful attention so that the complete methodology for measurement and marking is constantly being verified and validated. Despite the fact that there are technicians responsible for data collection who are fully trained in the methodology, a copy of the measurement protocol and its specifications should be available for consultation in the field.

It is important to remember that the period for plot measurement (completion of the first census or re-censuses) should be defined by the analysis of a series of rainfall in the region under study so that the measurements can always be done at the same time of year, that is, in the month that has the least amount of precipitation. This strategy seeks to take advantage of the best transportation logistics, generally by ground, and to avoid the influence of rain in the diameter measurements since tree bark can become saturated with water, thus affecting/falsifying growth data.

For the individual measurement of trees, it is recommended that diametric tape (e.g. Diameter Tape – Forest Suppliers) be strictly used. The use of tapes that measure the circumference of individual trees, in order to later convert to diameter, increases estimation errors. The technician responsible for measurement should note, tree by tree, any loose bark, lichens, lianas or other factors that could impact diameter measurement. The technician cleans the measurement area by passing his/her hand along the trunk and then runs the diametric tape around it. Also responsible for worksheet data, the technician should seek to assist the one who is measuring the trees, primarily during the evaluation of large trees, in order to verify the correct position of the tape and to determine if there is anything between the tape and the tree.

During the annual re-censuses, the technician responsible for recording data in the worksheets should pay even greater attention to the data that are found to be divergent from the previous year's records, which could likely be due to an error in reading the diametric tape. If an error is found, the technician should ask for a re-measurement and a re-reading of the diameter for recording in the worksheet.

Another important activity undertaken during the re-censuses is an active search throughout the PMP for new individuals to be included in the sample (recruits) and individuals that no longer exhibit vegetative activity (dead). All of the new trees, palms and lianas that have met the inclusion criteria (DBH ≥ 10 cm) are included in the sample and the same marking methodology is followed. Individuals marked in the first census but which, during the re-census, did not exhibit vegetative activity or were not found after a detailed sweep of the plot, should be considered dead.

It is also possible that trees that had died in the previous year show activity through diametric growth or new growth. In this case, the processing worksheet should be modified, correcting the data recorded the previous year and including this individual once again in the sample since it was not actually dead

3. Analyzing data

3.1. Tabulation of data

For all field activities related to planning, implementation and monitoring of PMPs, there should be specific worksheets. The standardization of the entry of information that will be generated is of fundamental importance to guarantee the quality of the data. Each worksheet should include the following information at minimum:

- PMP name and abbreviation;
- Complete date when the collection was done;
- Names of each of the team members;
- Number of each individual;
- Registration number of the sub-plot to which each individual belongs;
- Data related to the POM and DBH;
- Pertinent observations.

Upon completion of the field work, all of the worksheets used should be digitized, scanned, and saved in a digital file and then stored in a dry, safe place. These procedures assure that the original worksheets can be consulted in the case of duplicate or conflicting information, when typing errors occur, or when mistakes are made in noting information in the field. After digitizing the worksheet data, the new worksheets should be printed and evaluated by pairs for accuracy, followed by the correction of any confirmed errors.

3.2. Spatial mapping

Spatial mapping of the individuals marked in the PMPs allows for the possibility of analyses of the distribution of species or guilds in the forest. For these analyses, indices of aggregation, such as Morisita [15] or McGuinness [16], can be used, thus defining the spatial distribution of the individuals as aggregate, random or regular. This knowledge is fundamental to ecological analyses as it facilitates an understanding of how a certain species uses available resources in the forest. While the aggregation factor can vary within a species, in different diametric classes, it shows how the life stages of an individual can change the way it uses an available resource.

For mapping, each individual should have its Cartesian coordinates X and Y measured in the PMP. The distances can be measured using a 50 meter measuring tape or a digital measuring stick. It is important that a compass always be used to support the measurements so that the distances are consistently taken in a straight line with respect to the position within each sub-plot. In the example below (Figure 4), the individual marked in the PMP has Cartesian coordinates of X = 56.2 meters and Y = 74.3 meters.

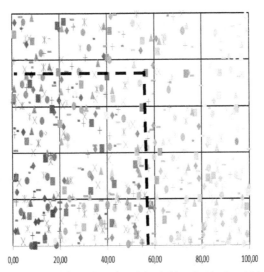

Figure 4. Example of the result of spatial mapping of the field individuals within the PMP at Rio Doce State Park – Minas Gerais, Brazil.

3.3. Estimates of biomass and carbon stocks

The estimates of aboveground live biomass and the resulting carbon stocks can be obtained using two key methods. The first, based on destructive sampling (direct method), involves cutting, drying and weighing separately (roots, trunk and leaves) all of the trees in a specific area. This technique becomes unviable in the case of monitoring since it can damage the sample over the life of the vegetation. The second method (indirect method) consists of estimating biomass and carbon stocks by measuring field variables without having to fall the tree. In this case, DBH data and/or total height of the trees (Ht) and/or specific density of the wood (p) are inserted into previously developed allometric equations in order to estimate the biomass and carbon stocks of the PMP.

Table 1 shows examples of allometric equations already developed and that can be used to calculate biomass. The selection of the best equation should be based on the objective of the project and on the questions to be answered. Allometric models that offer greater precision should be given preference [17].

Types	Allometric Equations	R^2
Wet Forest [18]	$EXP(-2.557 + 0.940 * LN(p * DBH^2 * Ht))$	0.99
Moist Forest [18]	$EXP(-2.977 * LN(p * DBH^2 * Ht))$	0.99
Dry Forest [18]	$EXP(-2.187 + 0.916 * LN(p * DBH^2 * Ht))$	0.99
Palms [19]	$\dfrac{EXP\left((5.7236 + 0.9285 * LN(DBH^2)) * 1.05001\right)}{10^3}$	0.82
Lianas [20]	$EXP\left(0.07 + 2.17 * (LN(DBH))\right)$	0.95
Amazon [21]	$EXP(-1.754 + 2.665 * LN(DBH))$	0.92
Amazon [21]	$EXP(-0.151 + 2.17 * LN(DBH))$	0.90
Tree ferns [22]	1266340/(1 - (279228↓ EXP(↑↓↓↕↕↕↕↕)))	0.00
Wet Forest [23]	$EXP(21.297 - 6.953(DBH) + 0.74(DBH^2))$	0.91

[18] – Chave et. al. (2005);
[19] – Nascimento & Laurance (2002);
[20] – Gerwing & Farias (2000);
[21] – Higuchi et. al. (1998);
[22] – Tiepolo et al. (2002);
[23] – Brown et al. 1997.

Table 1. Example of allometric equations used to estimate the aboveground biomass (kg) of trees, palms and lianas in different tropical forest types. DBH – Diameter at breast height; Ht – Total height; and p – Wood mean density g/m³.

In order to conduct accurate comparisons with other areas or to serve as a potential indicator of carbon stocks for a specific region, simpler allometric equations with only one variable – DBH can be used [17]. In this case, it is not necessary to collect data related to the height or wood density of individuals, resulting in the inventory being completed much faster. An important detail regarding the selection of the equation is that the results for some are fresh biomass data, while for others they are dry biomass data, and still others provide results as carbon quantity.

As previously mentioned, the ideal would be to use an allometric model that provides the highest degree of confidence. The best model has to explain most of the variation in the data or has the lowest AIC (Akaike Information Criterion). In the most cases, equations that use multiple entries with 3 variables per individual (DBH, Ht and p) are better. DBH data are easily collected as previously outlined. The data related to tree height are generally complicated to collect due to error associated with height estimations, in addition to the need for greater time in the field, which results in inventories having higher costs. In order to optimize this work, an estimate of tree height can be used by creating an allometric equation adjusted by the diametric and height measurements of a specific number of trees in the plot (Figure 4). This requires the collection of height data for a certain part of the plot. These data should be collected with the greatest precision possible, using cords, ladder or equipment such as a rangefinder. It is recommended that height be measured for a random sample of 20% of the individuals of a PMP in order to later relate them to the diameters, producing an equation for site-specific heights (Figure 4). In order to collect data for specific wood density, there are some protocols for extracting and obtaining these values for each tree in a PMP. With a view to obtaining economies of time and project resources, existing databases can be used, for example, Global Wood Density Database [24, 25], which makes available a series of wood density values for species that exist in almost every part of the world.

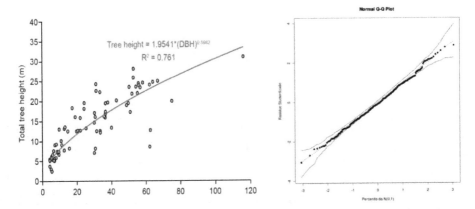

Figure 5. Examples of the development of a site-specific equation for the calculation of height using tree diameters [17] and of equation adjustment using the observance of the normality of residues.

With biomass calculated, many different possibilities for analysis become available. For example, comparisons of biomass can be done between primary and secondary areas, between one year and another, and total biomass can be calculated for the PMP and extrapolated to large forested areas of the same typology. In addition to comparing the relative data to the average annual increment of biomass (or of carbon) of a PMP, analyses of the change in biomass between years can also be conducted. This can be achieved by subtracting the biomass in year one from the biomass in year 0, remembering that this biomass value should include the biomass of recruits in year 1 while the biomass of individuals considered to be dead is subtracted. Another factor that can be considered is the number of days between each census in order to standardize the calculations for a specific period. For example, for 1 year, the following equation would be used:

$$\left(\frac{AGBt2-AGBt1}{DTt2-DTt1}\right) * 365; \tag{1}$$

where AGBt2 refers to biomass in year 2, and AGBt1 to biomass in year 1. DTt2 refers to the date the census was taken in year 2, and DTt1 to the date the census was taken in year 1 (D. Clark personal communication).

Table 2 shows the data for aboveground biomass (AGB) for different neotropical forest sites (adapted by Alves, 2010 [26]). The highest values were primarily found in the Brazilian Amazon (Manaus and Santarém).

Sites	AGB (Mg.ha⁻¹)	Reference
Submontane moist semideciduous secondary forest, Marliéria, Brazil	92.0	[27] Metzker et al. (2011)
Submontane moist semideciduous secondary forest, Marliéria, Brazil	107.0	[27] Metzker et al. (2011)
Lowland seasonally dry forest, Mexico	109.0	[28] Vargas et al. (2008)
Lowland wet forest, La Selva, Costa Rica	148.7	[11] Clark and Clark (2000)
Seasonally flooded forest (Restinga), Ubatuba, Brazil	151.0	[26] Alves et al. (2010)
Montane wet forest, Venezuela	157.0	[29] Delaney et al. (1997)
Montane moist forest, Venezuela	173.0	[29] Delaney et al. (1997)
Submontane moist semideciduous primary forest, Marliéria, Brazil	174.0	[27] Metzker et al. (2011)
Lowland moist forest, Venezuela	179.0	[29] Delaney et al. (1997)
Lowland moist forest, BCI, Panama	179.1	[30] DeWalt and Chave (2004)
Submontane moist semideciduous primary forest, Marliéria, Brazil	179.8	[27] Metzker et al. (2011)
Lowland moist forest, Ubatuba, Brazil	198.4	[26] Alves et al. (2010)
Submontane moist semideciduous primary forest, Marliéria, Brazil	201.0	[27] Metzker et al. (2011)
Lowland wet fores, La Selva, Costa Rica	203.2	[30] DeWalt and Chave (2004)
Lowland forests, SW Amazonia (Bolivia, Peru)	206.7	[31] Baker et al. (2004)
Lowland forests, NW Amazonia (Peru, Ecuador)	220.8	[31] Baker et al. (2004)
Submontane semideciduous forest, La Chonta, Bolivia	236.6	[32] Broadbent et al. (2008)
Submontane moist forest, Ubatuba, Brazil	239.3	[26] Alves et al. (2010)
Lowland wet forest, Manaus, Brazill	240.2	[30] DeWalt and Chave (2004)
Lowland moist forest, Rio Branco, Brazil	244.1	[33] Vieira et al. (2004)
Lowland moist forest, BCI, Panama	260.2	[34] Chave et al. (2003)
Montane moist forest, Ubatuba, Brazil	262.7	[26] Alves et al. (2010)

Lowland forests, Central & Eastern Amazonia (Brazil)	277.5	[31] Baker et al. (2004)
Lowland moist forest, Santarem, Brazil	281.2	[33] Vieira et al. (2004)
Lowland semideciduous forest, Roraima, Brazil	292.1	[35] Nascimento et al. (2007)
Lowland moist forest, Santarem, Brazil	294.8	[36] Rice et al. (2004)
Lowland moist forest, Santarem, Brazil	298.0	[37] Pyle et al. (2008)
Lowland moist forest, Rondonia, Brazil	306.8	[38] Cummings et al., 2002
Lowland wet forest, Manaus, Brazil	307.6	[39] Castilho et al. (2006)
Lowland wet forest, Nouragues, French Guiana	317.0	[40] Chave et al. (2001)
Lowland wet forest, Manaus, Brazil	325.5	[19] Nascimento and Laurance (2002)
Lowland moist Cocha Cashu, Peru	332.8	[30] DeWalt and Chave (2004)
Lowland wet forest, Manaus, Brazil	334.0	[37] Pyle et al. (2008)
Lowland semideciduous forest, Linhares, Brazil	334.5	[41] Rolim et al. (2005)
Lowland wet forest, Manaus, Brazil	360.2	[33] Vieira et al. (2004)

Table 2. Estimates of aboveground biomass in different forest typologies on neotropical sites. Adapted by (Alves et al. 2010 [26]). AGB data (Mg.ha[-1]).

3.4. Recruitment and mortality rates

Calculations of the annual rates of Recruitment (Eq. 2) and Mortality (Eq. 3) can be done using the equations by Sheil and Mail [42]. These rates are an excellent indicator of forest dynamics, providing a solid understanding of forest behaviour as it is affected by seasonal events causing variations in water availability, or by extreme climatic events or to conduct multiple comparisons. Since, in reality, everything depends on the proposed objective, forest dynamics can be compared, for example, between those individuals that belong to the higher diametric classes and those who belong to the lower, or the behaviour between different species, among others.

$$R = \left[\left(\left(\frac{No + Nr}{No} \right)^{\left(\frac{1}{t} \right)} \right) - 1 \right] * 100 \tag{1}$$

$$M = 1 - \left[\left(\left(\frac{No-Nm}{No}\right)^{\left(\frac{1}{t}\right)}\right)\right] * 100; \tag{2}$$

where: No equals the number of individuals at time 0; Nm is the number of dead individuals between the interval; and Nr is the number of individuals recruited in the same time interval (t).

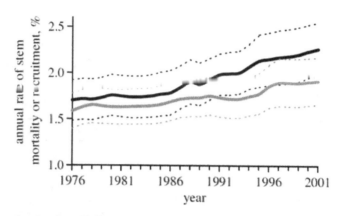

Figure 6. Example taken from Phillips et al., (2008) [43] referring to the analysis of mortality rates (grey lines) and recruitment rates (black lines), using a monitoring time period of 25 years. Solid lines are means and dotted lines are 95% CIs.

4. The valuation of tropical forests

In this section, we will explain how to assign value to carbon stock estimates taken from collected data on PMPs. We will also discuss issues regarding the Payment for Environmental Services (PES), which is provided by tropical forests that are connected with major international protocols and signed agreements.

4.1. Assigning value

Forest conservation strategies to be effective, local communities must first be significantly involved and they must believe in the importance of biodiversity to guarantee quality of life. These communities are the key to a conservationist network. The second step is to invest financially in these initiatives. The project should clearly demonstrate that forest conservation efforts are more economical lucrative when compared with the opportunity costs of using the soil in a given region, for example, for cattle-raising. Therefore, investing in the protection of biodiversity in order to encourage the social and economic development of local communities is one of the best long-term conservation strategies for biodiversity and the ecosystemic services it generates.

One of the difficulties in assigning value to biodiversity and the services it offers is how to specifically quantify this value. First, the value of its natural attributes is immeasurable, such as the services offered by bees when pollinating plantations throughout the world or the atmospheric regulation offered by forests (see, [44]). Thus, the carbon valuation and commercialization market has an advantage, since the prices per ton are already known by the market. Despite being affected by countries' economic changes, a ton of carbon (sequestered or saved) has its own regulations derived from international agreements, such as the Kyoto Protocol or by mechanisms such as the CDM (Clean Development Mechanism) and REDD (Reducing Emissions from Deforestation and Forest Degradation). Therefore, projects that seek to assign economic value to environmental services can include "carbon valuation" as a more precise indicator of the technical reliability of the project.

Forest projects began to participate in the global carbon credit market when companies partnered in order to preserve forests and plant trees with the goal of neutralizing their greenhouse gas emissions [3]. Due to the initial difficulty of negotiating these credits within a regulated market (compliance market), many of these initiatives looked for the voluntary market [3] and other financial transactions that could neutralize their emissions by trees capturing carbon. These new mechanisms opened the door for a wide variety of carbon projects that include voluntary initiatives as payment for the recovery of degraded areas as a means of neutralizing emissions and even responsibility for conserving existing forest areas.

4.2. Development of public policies

These widely diverse ongoing carbon projects have one objective in common: to take advantage of existing market mechanisms in order to assign economic values to rainforests. Today, the REDD+ mechanisms is considered one of the most interesting since it focuses on creating an institutional structure and economic incentives required for developing countries to substantially reduce their CO_2 emissions resulting from deforestation and forest degradation [45].

A practical example of implementing public policies connected to carbon projects is the program called Bolsa Floresta (Forest Fund), created by the state of Amazonas through Law no. 3135 on 05/06/2007. Through this initiative, the Government pays R$50 (~USD $30) per month to registered families who live in State Conservation Units and who have signed a collective agreement to stop deforestation [45]. In the state of Minas Gerais, the Government created an initiative called Bolsa Verde (Green Fund) (Law no. 17.127 in 2008), whose objective is to help conserve native vegetation cover in the State by paying property owners for environmental services if they already preserve or are committed to restoring native vegetation on their properties. In this case, the financial incentive is relative to the size of the protected area, which is a priority for family farms and rural producers. Thus, the REDD+ has a comprehensive rural planning strategy that values rainforests and their recovery, as well as supporting the sustainable development of rural livelihoods [45] and facilitating true socio-environmental gains.

For all of these initiatives works, there must also be reliable data on existing carbon stocks to serve as a baseline for the projects. Permanent Monitoring Plots are technically considered to be the best way to obtain these data. For forest recovery projects, where it is not possible to implement PMPs, they can be implemented using adjacent areas or existing data can even be used to extrapolate biomass values. During the monitoring of carbon projects, the random distribution of PMPs serves as a statistically equivalent sample area for forest recovery monitoring. As an example of other monitoring sites using a standardized methodology we can cite the TEAM network (http://teamnetwork.org/) which has more than 15 monitoring sites in tropical forests. In Brazil we can cite two of these sites: Manaus, Caxiuanã, which have 05 PMPs each. Another success case in the monitoring area is the LBA project (http://lba.inpa.gov.br/lba/), which has a vast network of PMPs in the Amazon forest, that in ten years been able to train more than 500 masters and doctors in Brazil, publishing ~1000 articles in specialized journals

Regardless of the type of project or the mechanism that is used to implement it, projects that use a ton of carbon (sequestered or saved) as the base, guarantee the long-term presence of these stocks in nature. But most importantly, these projects require the assured quality of the data they propose to collect. These data should have Measurement, Reporting and Verification (MRV) to guarantee the technical quality of the project (e.g. see the Standard CCBA – Climate Community and VCS - Voluntary Carbon Standard). In order to guarantee viability, these projects should also have local community involvement as a goal, whether in the implementation phase or during monitoring, in order to facilitate the improvement of quality of life and the resulting socio-environmental gains. In addition to facilitating the socio-environmental benefits already outlined, the implementation of local PMPs has a powerful differential: calibrating the calculation of international methodologies with highly reliable data, collected locally and using a standardized methodology [27].

Author details

Thiago Metzker
Programa de Pós-Graduação em Ecologia, Conservação e Manejo da Vida Silvestre, Brazil

Tereza C. Spósito
Departamento de Botânica, Universidade Federal de Minas Gerais (UFMG),
CP 486, CEP 31270-970, Belo Horizonte, Brazil

Britaldo S. Filho
Departamento de Geociências, UFMG, Belo Horizonte, Brazil

Jorge A. Ahumada
Tropical Ecology Assessment and Monitoring Network, Science and Knowledge Division,
Conservation International, Arlington, VA, USA

Queila S. Garcia
Departamento de Botânica, Universidade Federal de Minas Gerais (UFMG),
CP 486, CEP 31270-970, Belo Horizonte, Brazil

Acknowledgement

Support for this research was received through FAPEMIG (Fundação de Amparo à Pesquisa no Estado de Minas Gerais, Process APQ-02183-09), PELD (Long Term Ecological Research – CNPq, Process 520031/98-9), the Tropical Ecology Assessment and Monitoring (TEAM) Network, a collaboration between Conservation International, the Missouri Botanical Garden, the Smithsonian Institution, and the Wildlife Conservation Society, and partially funded by these institutions, the Gordon and Betty Moore Foundation, and other donors, and USF&WS. We thank Edgar Paiva for the illustrations in this chapter, Orbifish Global Solutions for linguistic review and the whole community of Rio Doce State Park. We thank UFMG, ECMVS and IEF (State Forestry Institute) for logistical support. T.M. received a Doctor fellowship from CAPES (Brazil) and Q.S.G. received a scholarship from CNPq (Brazil).

5. References

[1] FAO. Global Forest Resoucers Assessment (2005). Progress towards sustainable forest management. FAO Forestry Paper 147: Food and Agriculture Organization of the United Nations. 350 p.

[2] Keeling, H.C. & Phillips, O.L. (2007) The global relationship between forest productivity and biomass. Global Ecol. Biogeogr. Vol. 16: 618–631.

[3] MMA (2011). Pagamento por Serviços Ambientais na Mata Atlântica: lições apreendida e desafios. Guedes, F.B. & Seehusen, S.E. (Eds.). Série Biodiversidade 42. 276 p.

[4] Corona, P. et al. (2011). Forest Ecology and Management Contribution of large-scale forest inventories to biodiversity assessment and monitoring. Forest Ecology and Management, Vol. 262: n 11, p. 2061-2069.

[5] Lund, H.G. et al. (1998). Plots, pixels, and partnerships: potential for modeling, mapping and monitoring biodiversity. In: Forest Biodiversity Research, Monitoring and Modeling: Conceptual Background and Old World Case Studies, F. Dallmeier, J.A. Comiskey, (Eds.). Man and the Biosphere Series. Vol. 20: UNESCO & The Parthenon Publishing Group. Carnforth, Lancashire, UK.

[6] Andelman, S.J. and Willig, M. R. (2004). Networks by Design: A Revolution in Ecology. Science. Vol: 305. 2004.

[7] TEAM Network (2010). Vegetation Protocol Implementation Manual. Tropical Ecology, Assessment and Monitoring Network, Science and Knowledge Division, Conservation International. Vol. 1.5.1: 75 p.

[8] Malhi, Y. et al. (2002). An international network to understand the biomass and dynamics of Amazonian forests (RAINFOR). Journal of Vegetation Science. Vol 13: 439-450.

[9] Condit, R. (1998). Tropical Forest Census Plots. Springer-Verlag, Berlin, and R. G. Landes Company, Georgetown, Texas.

[10] TEAM Network. (2011). TEAM Network Sampling Design Guidelines. Tropical Ecology, Assessment and Monitoring Network, Science and Knowledge Division, Conservation International, Arlington, VA, USA.

[11] Clark, D.B. & Clark, D.A. (2000). Landscape-scale variation in forest structure and biomass in a tropical rain forest. Forest Ecology and Management, Vol: 137. 185–198.

[12] Clark, D.A. (2002). Are tropical forests an important carbon sink? Reanalysis of the long-term plot data. Ecological Applications. Vol 12: 3–7.

[13] Chave, J. et al. (2004). Error propagation and scaling for tropical forest biomass estimates. Phil. Trans. R. Soc. Lond. B. Vol 359: 409–420.

[14] Schaaf L. B. et al. (2205). Incremento diamétrico e em área basal no período 1979-2000 de espécies arbóreas de uma floresta ombrófila mista localizada no sul do Paraná Florestu. Vol 35?

[15] Morisita, M. (1959). Measuring of the dispersion of individuals and analysis of the distributional patterns. Men. Fac.Sci. Kyushi Univ., Ser. E (Biol.) Vol. 2: n. 4, p. 215 – 235.

[16] McGuinnes, W.G. (1934). The relationship between frequency index and abundance as applied to plant populations in a semi-arid region. Ecology. Washington. Vol 16: p. 263-282.

[17] Vieira, S.A. (2008). Estimation of biomass and carbon stocks: the case of the Atlantic Forest. Biota Neotrop Journal. Vol. 8: n 2. ISSN 1676-0603. Available at: http://www.biotaneotropica.org.br/v8n2/pt/fullpaper?bn00108022008+en

[18] Chave, J. et al. (2005) Tree allometry and improved estimation of carbon stocks and balance in tropical forests. Oecologia. Vol 145: 87–99.

[19] Nascimento, H. E. & Laurance, W.F. (2002). Total aboveground biomass in central Amazonian rainforest: a landscape-scale study. For. Ecol. Manage. Vol 168: 311–321.

[20] Gerwing, J. J. & Farias, D. L. (2000). Integrating liana abundance and forest stature into an estimate of aboveground biomass for an eastern Amazonian forest. J. Trop. Ecol. Vol 16: 327–336.

[21] Higuchi, N. et al. (1998). Biomassa da parte aérea da vegetação da floresta tropical úmida de terra firme da Amazônia Brasileira. Acta Amazônica. Vol 28(2): 153-166.

[22] Tiepolo, G. et al. (2002). Measuring and monitoring carbon stocks at the Guaraquec, aba Climate Action Projetct, Paraná, Brasil. In: International Symposium on Forest Carbon Sequestration and Monitoring. Extension Series Taiwan Forestry Research Institute. Available at: http://www.spvs.org.br/download/monitoramento ingles.pdf, pp. 98– 115.

[23] Brown, S. (1997). Estimating Biomass and Biomass Change of Tropical Forests: a Primer. For the food and agriculture organization of the united nations. Rome, FAO Forestry Paper 134.

[24] Chave J. et al. (2009). Towards a worldwide wood economics spectrum. Ecology Letters. Vol 12: 351-366.

[25] Zanne A. E. et al. (2009). Data from: Towards a worldwide wood economics spectrum. Dryad Digital Repository.

[26] Alves, L. F. (2010). Forest Ecology and Management Forest structure and live aboveground biomass variation along an elevational gradient of tropical Atlantic moist forest (Brazil). Forest Ecology and Management. Vol 260(5): 679-691. doi:10.1016/j.foreco.2010.05.023. http://dx.doi.org/10.1016/j.foreco.2010.05.023.

[27] Metzker, T. et al. (2011). Forest dynamics and carbon stocks in Rio Doce State Park – an Atlantic rainforest hotspot. Current Science. Vol 100(12): 2093-2098. Available at: http://cs-test.ias.ac.in/cs/Volumes/100/12/1855.pdf

[28] Vargas, E. et al. (2008). Biomass and carbon accumulation in a fire chronosequence of a seasonally dry tropical forest. Glob. Change Biol. Vol 14, 109–124.

[29] Delaney, M. et al. (1997). The distribution of organic carbon in major components of forests located in five life zones in Venezuela. J. Trop. Ecol. Vol 13, 697–708.

[30] DeWalt, S.J. & Chave, J., 2004. Structure and biomass of four lowland Neotropical forests. Biotropica. Vol 36: 7–19.

[31] Baker, T.R. et al. (2004). Variation in wood density determines spatial patterns in Amazonian forest biomass. Glob. Change Biol. Vol 10: 545–562.

[32] Broadbent, E. et al. (2008). Spatial partitioning of biomass and diversity in a lowland Bolivian forest: linking field and remote sensing measurements. Forest Ecol. Manag. Vol 255: 2602–2616.

[33] Vieira, S.A., et al. (2004). Forest structure and carbon dynamics in Amazonian tropical rain forests. Oecologia. Vol 140: 468–479.

[34] Chave, J. et al. (2003). Spatial and temporal variation of biomass in a tropical forest: results from a large census plot in Panama. J. Ecol. Vol 91: 240–252.

[35] Nascimento, M.T. et al. (2007). Above-ground biomass changes over an 11-year period in an Amazon monodominant forest and two other lowland forests. Plant Ecol. Vol 192: 181–191.

[36] Rice, A.H. et al. (2004). Carbon balance and vegetation dynamics in an old-growth Amazonian forest. Ecol. Appl. Vol 14: 55–S71.

[37] Pyle, E.H. et al. (2008). Dynamics of carbon, biomass, and structure in two Amazonian forests. J. Geophys. Res. Vol 113: G00B08.

[38] Cummings, D.L. et al. (2002). Aboveground biomass and structure of rainforests in the southwestern Brazilian Amazon. Forest Ecol. Manag. Vol 163: 293–307.

[39] Castilho, C.V.et al. (2006). Variation in aboveground tree live biomass in a central Amazonian Forest: effect of soil and topography. Forest Ecol. Manag. Vol 234: 85–96.

[40] Chave, J. et al. (2001). Estimation of biomass in a Neotropical forest of French Guiana: spatial and temporal variability. J. Trop. Ecol. Vol 17: 79–96.

[41] Rolim, S.G. et al. (2005). Biomass change in an Atlantic tropical moist forest: the ENSO effect in permanent sample plots over a 22-year period. Oecologia. Vol 142: 238–246.

[42] Sheil, D. & May, R.M. (1996). Mortality and recruitment rate evaluations in heterogeneous tropical forests. Journal of Ecology. Vol 84: 91–100.

[43] Phillips, O. L. et al. (2008). The changing Amazon forest. Philosophical Transactions of the Royal Society B Biological Sciences. Vol 363: 1819-1827. Available at: http://www.pubmedcentral.nih.gov/articlerender.fcgi?artid=2374914&tool=pmcentrez&rendertype=abstract.

[44] Artaxo, P. et al. (2005). Química atmosférica na Amazônia: a floresta e as emissões de queimadas controlando a composição da atmosfera amazônica. Acta Amazônica. Vol 35:185-196.

[45] Soares Filho, B. et al. (2012). Challenges for Low-Carbon Agriculture and Forest Conservation in Brazil. IDB Publications 64798, Inter-American Development Bank.

Chapter 7

Decreased Epiphytic Bryophyte Diversity on Mt. Odaigahara, Japan: Causes and Implications

Yoshitaka Oishi

Additional information is available at the end of the chapter

1. Introduction

When wildlife populations grow excessively, they affect other flora and fauna within their ecosystems (Fuller & Gill, 2001; Pellerin et al., 2006; Rooney, 2001; Schütz et al., 2003; Stewart & Burrows, 1989; Stockton et al., 2005; Takatsuki, 2009; Webster et al., 2005). For example, the recent increase in the sika deer population in Japan has led to the degradation of ecosystems in many areas. From 1979 to 2002, the range of this species expanded by as much as 70% (Nakajima, 2007). Although stripping of bark, grazing on grass, and browsing on tree understories are normal foraging behaviors in deer, these activities in excess can cause severe damage. Excessive bark stripping causes wood decay, leading to a decline in the forest cover (Akashi & Nakashizuka, 1999; Miquelle & van Ballenberghe, 1989; Yokoyama et al., 2001), and excessive browsing and/or grazing may alter the structure and composition of vegetation on the forest floor (Kumar et al., 2006; Rooney & Waller, 2003; Schütz et al., 2003; Stockton et al., 2005; Webster et al., 2005). These environmental changes indirectly affect other organisms in the forest ecosystem (Allombert et al., 2005; Feber et al., 2001; Flowerdew & Ellwood, 2001; Rooney, 2001).

To protect forest vegetation from further damage by the increased sika deer population, protective management, for example, wrapping tree trunks in wire mesh, have been implemented in addition to deer population control via culling and the erection of deer-proof fences (Ministry of the Environment-Kinki Regional Environment Office, 2009; Takatsuki, 2009). However, although these measures are effective for the protection of vegetation, they sometimes negatively affect other organisms.

This chapter describes the effects of protective management activities on epiphytic diversity at Mt. Odaigahara in central Japan, which is a hotspot for bryophyte diversity. It also discusses the best practices for biodiversity conservation in this scenario on the basis of a previously published article (Oishi, 2011).

2. Study site

2.1. Location and characteristics of the study site

Mt. Odaigahara (34°N, 136°E; altitude, ca. 1,500 m) is located in Yoshino Kumano National Park, which is in the southeastern part of the Nara Prefecture in Japan (Fig. 1). The climate in this region is relatively mild (annual mean temperature, 5.7 °C), with high levels of precipitation (annual mean precipitation, 4,500 mm; Nara Local Meteorological Observatory 1997). The vegetation on Mt. Odaigahara is classified into 2 main types: (1) the dominant tree species on the eastern part of the mountain is *Picea jezoensis* (Sieb. et Zucc.) Carriere var. *hondoensis* (Mayr) Rehder, and (2) those on the western part are *Fagus crenula* Blume and *Abies homolepis* Sieb. et Zucc (Ide & Kameyama, 1972).

2.2. Deer population

The population density of sika deer in Mt. Odaigahara has rapidly increased from the 1960s to the 1990s. To be specific, it has increased from approximately 12.0–22.2 individuals per square kilometer in the 1980s to 17.5–39.5 individuals per square kilometer in the 1990s (Ando & Goda, 2009). This increase led to serious damage to the forest vegetation in this

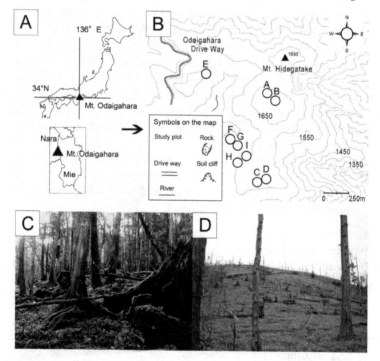

A; Location of Mt. Odaigahara in Japan. B: Location of study plots. C and D: Views of forests in Mt. Odaigahara. Photo C shows a forest of *P. jezoensis* var. *hondoensis* trees, whereas D shows a heavily declined forest.

Figure 1. Mt. Odaigahara and study plots

mountain region; for example, extensive bark stripping by the deer resulted in the dieback of damaged trees, and excessive browsing/grazing led to the loss of vegetation on the forest floor (Fig. 2). The Ministry of the Environment initiated a forest protection program in 1986 to conserve the forest ecosystem in this region. This program was executed in a part (ca. 703 hectare) of Mt. Odaigahara (Ministry of the Environment, Kinki Regional Environment Office, 2009). To prevent bark stripping by the deer, the trunks of around 32,500 trees were wrapped with wire mesh composed of zinc-coated galvanized iron (Ministry of the Environment, Kinki Regional Environment Office, 2009) (Fig. 3).

The photo on the left shows *Sasa nipponica* Makino et Shibata. browsed by sika deer. In this photo, the upper parts of the plants were browsed by sika deer. The right photo shows a deer fence and the effects of protection from browsing by sika deer: the height of the plants within the deer fence (back) is greater than that of the plants outside the fence.

Figure 2. Deer fence and the influence of browsing by deer on vegetation

The middle part of the tree trunk that does not have wire mesh protection has been debarked by deer.

Figure 3. Examples of tree trunks without (left) and with (right) wire mesh protection

2.3. Bryophyte diversity

In addition to the rapidly increasing population of sika deer on Mt. Odaigahara, this region is recognized for its bryophyte diversity. In fact, Mt. Odaigahara is home to approximately 30% (>620 species) of the bryophytes in Japan (Doei, 1988), including several nationally endangered species that are listed in the Red Data Book of Japan (e.g., *Iwatsukia jishibae* (Steph.) N. Kitag). The rich diversity of epiphytic bryophytes in this region is attributed to the high humidity of this region (Doei, 1988) (Fig. 4). The major species on Mt. Odaigahara include *Pogonatum japonicum* Sull. & Lesq., *Dicranum japonicum* Mitt., *Hylocomium splendens* (Hedw.) Schimp., and *Pleurozium schreberi* (Brid.) Mitt., which grow on the forest floor; *Heterophyllium affine* (Hook.) M. Fleisch, *Bazzania yoshinagana* (Steph.) S. Hatt., *Mylia verrucosa* Lindb., and *Scapania ampullata* Steph., which grow at the base of trees; and *Pterobryon arbuscula* Mitt., *Hypnum tristo-viride* (Broth.) Paris, and *Bazzania denudata* (Torr. ex Lindenb.) Trevis., which grow on tree trunks (Fig. 5). Bryophytes contribute to species diversity as well as the ecological integrity of Mt. Odaigahara because they function as microhabitats for the seedbeds of vascular plants and are involved in rainfall interception and nutrient cycling (Coxson, 1991; Nadkarni, 1984; Nakamura, 1997; Pypker et al., 2006) (Fig. 6).

2.4. Changes in bryophyte diversity and the deer population since the 1960s

In the 1980s, epiphytic bryophyte flora in *P. jezoensis* var. *hondoensis* were surveyed in 2 parts of Mt. Odaigahara—Masakitoge and Masakigahara. These areas were dense forests in the 1960s, but by 2008, they had become open forests because the changes in the environmental conditions after a severe typhoon in 1959 gradually resulted in dieback. Because many trees were blown down by the typhoon, the light conditions in the forests improved, and *S. nipponica* grew vigorously on the forest floor. This expansion resulted in an increase in the population of sika deer, which in turn resulted in dieback due to debarking.

The lower parts of many tree trunks (left) and fallen logs (right) are extensively covered with epiphytic bryophytes because of the high humidity.

Figure 4. Epiphytic bryophytes in Mt. Odaigahara

These species frequently occurred in the study plots: A, *D. japonicum*; B, *P. arbuscula*; C, *H. affine*; D, *H. tristo-viride*; E, *H. splendens*; F, *M. verrucosa*

Figure 5. Major species in the study plots

In 2008, we surveyed the epiphytic bryophyte flora in almost the same places as those examined in a previous study by using 20 × 20 m quadrants (plots A, B, C, and D in Fig. 1), and we examined the changes in these areas that had occurred over in the last 30 years. Table 1 summarizes the environmental conditions in these plots.

The species richness of individual *P. jezoensis* var. *hondoensis* trees decreased over 30 years from 18.0 ± 3.5 to 5.7 ± 3.4 in Masakitoge and from 18.0 to 7.5 ± 5.3 in Masakigahara (mean or mean $\pm$ SD) (Fig. 7). Thus, in direct contrast to the increasing deer population, epiphytic bryophyte diversity significantly declined over time.

Possible reasons for the decrease in bryophyte diversity are that (1) the decline in the forest cover indirectly affected bryophyte diversity because of the changes in the environmental conditions (e.g., air humidity), and (2) the protection of trees using wire mesh directly affected bryophyte diversity because of metal pollution.

To determine the reasons for the decline in bryophyte diversity, we examined the correlation between the diversity of epiphytic bryophytes and environmental variables, including wire mesh protection

Figure 6. Bryophyte function in the ecosystem

Bryophytes provide safe microhabitats for the seedbeds of vascular plants (left). B: Bryophytes absorb water from rain drops and mist and therefore function in water storage in forests (right).

Plot	Year	No. of *Picea jezoensis* var. *hondoensis* trees surveyed		DBH (mean $\pm$ S.D.)
		Total	With wire mesh	
Masakitoge	1980s	2	0	23.7
	2008	10	8	23.8 ± 4.2
Masakigahara	1980s	13	0	34.4 ± 7.6
	2008	10	9	25.0 ± 11.2

Table 1. Summary of the characteristics of the study plots sampled for comparing the changes in the species richness from the 1980s to 2008

The bars represent the mean value of species richness and epiphyte cover on a single tree, and the error bars represent the corresponding standard deviations.

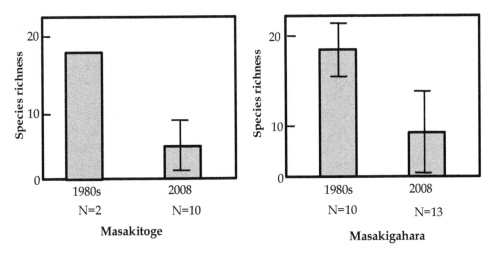

Figure 7. Species richness of epiphytic bryophytes on a single *P. jezoensis* var. *hondoensis* tree in the 1980s and 2008

3. Bryophyte diversity and environmental variables

3.1. Major environmental factors influencing bryophyte diversity

3.1.1. Site selection

A preliminary survey was conducted to identify plots of forest that were dominated by *P. jezoensis* var. *hondoensis* trees, including those with and without a protective wire mesh. In total, 9 plots (each 20 × 20 m in size) were selected (Fig. 1 A–I; Table 2) to examine the influence of wire mesh protection on epiphytic bryophyte diversity. Plots A–D were identical to those mentioned in section 2.4. We observed that the tree trunks were completely wrapped with wire mesh, from the ground up to a height of 150–180 cm. The mesh was composed of zinc-coated galvanized iron, a commonly used material for wire meshes (Japan Society of Corrosion Engineering, 2000). In each plot, the tree density (m²/plot) was measured on the basis of the total basal area of the trunks. Further, the fraction of the trunk area of the *P. jezoensis* var. *hondoensis* trees that had been debarked by sika deer was also recorded in each plot.

3.1.2. Bryophyte sampling

The epiphytic bryophyte flora on the trunks of *P. jezoensis* var. *hondoensis* trees in the study plots were surveyed from October to November 2008. The bryophyte species covering the tree trunks from ground level to a height of 1.5 m were examined.

Bryophyte nomenclature followed that reported by Iwatsuki (2001). The proportion of bryophyte cover, as a percentage of the total available bark area being investigated, was divided into 6 levels: 1 (<1%), 2 (≥1% to <10%), 3 (≥10% to <25%), 4 (≥25% to <50%), 5 (≥50% to <75%), and 6 (≥75%).

3.1.3. Analysis of the correlation between bryophyte diversity and environmental variables

A simple generalized linear model (GLM) using R software for Windows 2.11.0 (R Development Core Team, 2010) was used to identify the correlations between species richness and the bryophyte cover with respect to environmental variables. To identify the most parsimonious model, we performed automated stepwise model selection using the Akaike information criterion (AIC) using the minimum AIC as the best-fit estimator. Bryophytes that had been identified only up to the genus level were not included in the calculation of species richness if any species of that genus was sampled. The environmental variables used in the GLMs were tree density, host tree diameter at breast height (DBH), percentage of delimited area, and percentage of tree trunks with wire mesh protection.

3.1.4. Results & discussion

Associated with the 110 tree trunks in the sampling plots, 68 species were identified in the bryophyte flora survey: 29 mosses and 39 liverworts (Appendix). Fig. 8 shows the species richness and cover in the study plots. The species richness on a single tree ranged from no species to 34 species (mean = 9.1, SD ± 9.0), while the bryophyte cover ranged from 0 to level 5 (mean = 2.0, SD ± 1.8).

The GLMs constructed using the environmental variables are presented in Table 3. These models showed that the species richness and bryophyte cover were significantly correlated with DBH (height, 1.5 m) and tree density ($p < 0.01$) but negatively correlated with the presence of wire mesh protection ($p < 0.01$). The GLMs for species richness and bryophyte cover explained 70.1% and 80.4% of the variance, respectively ($p < 0.01$ for both models).

High tree density and host tree DBH have been suggested to be beneficial for bryophyte diversity, as they provide better microclimates, e.g., humid conditions (Hazell et al., 1998; Ojala et al., 2000; Thomas et al., 2001). Further, the species richness and bryophyte cover may be positively correlated with high host tree DBH because DBH is correlated with bark features (e.g., bark thickness and bark roughness) (Boudreault et al., 2008; Ojala et al. ,2000).

The results of this study raise the question of how wire mesh protection negatively affects bryophyte diversity. Considering that the galvanized iron, which is the primary component of the wire mesh, is coated with zinc, it is likely that the zinc affects the bryophytes. In the next section, we compare the zinc concentrations in bryophytes on trees with and without wire mesh protection.

3.2. Effect of wire mesh protection on bryophytes)

3.2.1. Bryophyte samples

To examine the influence of wire mesh protection on the bryophytes, inductively coupled plasma-mass spectrometry (ICP-MS) was used to compare the concentration of zinc in bryophyte samples, since zinc coats the wire mesh surface. For this evaluation, 2 species of bryophyte that are commonly found on the trunks of *P. jezoensis* var. *hondoensis* trees of this region, both with and without wire mesh, were sampled: *H.tristo-viride* and *S. ampliata*. For each species, 3 sets of samples each were collected from trees with and without wire mesh.

3.2.2. Analysis of zinc concentration

Dry samples (0.05–0.10 g) were placed in polytetrafluoroethylene vessels and weighed. Subsequently, 5 mL of nitric acid was added to the samples, and they were digested using a microwave system (MLS-1200 MEGA; Milestone General, Tokyo, Japan) before ICP-MS analysis. The samples were then analyzed using a 7500CX ICP-MS system (Agilent Technologies, Wilmington, DE, USA). Spectral interference was minimized or eliminated using the octopole reaction system, with helium as the reaction gas at a flow rate of 2.5 mL/min. The ICP-MS analysis was repeated twice for each sample, and the mean values were used in one-sided Student's t-test comparisons of the zinc concentration from bryophyte samples on tree trunks with and without wire mesh.

Plot	Altitude	Tree density (m²/plot)	No. of *Picea jezoensis* var. *hondoensis* trees	
			Total	With wire mesh
Plot A	1676	3.1	3	2
Plot B	1672	4.0	7	6
Plot C	1621	5.5	2	2
Plot D	1619	4.3	8	7
Plot E	1597	13.9	12	0
Plot F	1572	11.9	15	0
Plot G	1576	19.9	11	11
Plot H	1590	9.9	30	30
Plot I	1597	12.7	22	0

Table 2. Summary of the characteristics of the study plots sampled, including altitude, tree density, and number of trees surveyed

Variables	Cover			Species richness		
	coefficient s	t-value	p	coefficient s	t-value	p
Intercept	7.30	3.81	<0.01	2.26	7.30	<0.01
Tree density	3.51×10^{-4}	3.10	<0.01	6.30×10^{-5}	3.43	<0.01
Host tree DBH	2.01×10^{-1}	3.12	<0.01	2.29×10^{-2}	2.18	<0.05
Wire mesh	-1.43×10^{-1}	−14.2	<0.01	-3.19×10^{-2}	−19.5	<0.01
Adjusted R squared	0.701			0.804		

Table 3. Generalized liner models showing the association of species richness and bryophyte cover with environmental variables

The significance level of the coefficients and adjusted R² values are shown.

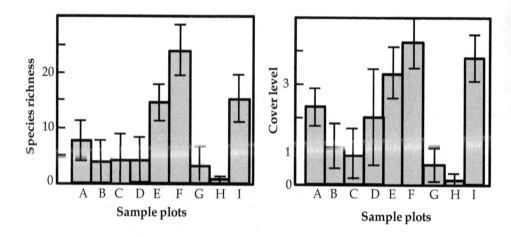

The bars represent the mean value of species richness and epiphyte cover on a single tree, and the error bars represent the corresponding standard deviations. Most tree trunks in plots A, B, C, D, G, and H had wire mesh protection.

Figure 8. Comparison of bryophyte species richness between trees with and without wire mesh protection

3.2.3. Results & discussion

ICP-MS analysis showed a significant 3- to 6-fold higher concentration of zinc in bryophytes inhabiting the bark of trees with wire mesh protection than in those without wire mesh protection (Fig. 9). Previous studies have shown that a considerable amount of zinc is leached from the zinc coating of galvanized iron by rain and dew (Harris, 1946; Seaward, 1974). Research has also shown that zinc is highly toxic to bryophytes (Tyler, 1990). Consequently, from the decreased diversity and increased zinc concentration of bryophytes on trees with wire mesh protection, it is reasonable to conclude that the loss of bryophyte cover and species richness has primarily occurred because of the toxicity of the zinc in the wire mesh. Additionally, other heavy metals in the wire mesh (e.g., iron) may affect bryophytes, with different heavy metals exerting varying levels of toxicity for bryophytes (Tyler, 1990).

4. Implications for biodiversity conservation

The results show that epiphytic bryophyte diversity is positively influenced by tree density and host tree DBH but negatively influenced by wire mesh protection, because of zinc toxicity (Fig. 10). The decline in bryophyte abundance and diversity on the lower parts of the tree trunks may be a cause for concern for biodiversity conservation on Mt. Odaigahara. This is because bryophytes contribute significantly to the species richness and biomass of tree trunks (Fritz, 2009; Lyons et al., 2000), as well as for ecosystem functions.

Furthermore, in addition to bryophytes, tree bark also provides important habitats for lichens and vascular epiphytes (Williams & Sillett, 2007). However, as heavy metals are toxic to these plants (Tyler et al., 1989), wire mesh protection may also contribute towards decreasing their levels of diversity and ecosystem functions. Unfortunately, considering that wire mesh protection is generally used against mammalian pests due to its direct effectiveness (Salmon et al. 2006; Vercauteren et al. 2006), this negative impact on bryophyte diversity may be widespread.

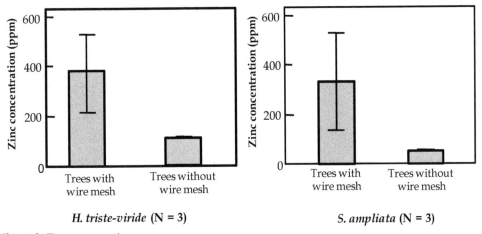

H. triste-viride (N = 3)   S. ampliata (N = 3)

Figure 9. Zinc concentration

The zinc concentration was significantly higher in bryophytes on trees with wire mesh than in those without wire mesh ($p < 0.05$; t-test).

Therefore, to establish best practices for biodiversity conservation that includes bryophytes, we should not only protect trees against bark stripping by deer but also focus on the materials used for protection. Alternative techniques for plant protection include the use of tree shelters in which trees are enclosed in plastic tubes (Ward et al., 2000) and forest enclosures using plastic mesh fencing (Vercauteren et al., 2006). However, these alternatives may also affect biodiversity conservation. For example, tree shelters decrease light transmission (Ward et al., 2000), which might alter the composition of bryophyte species on tree trunks. Further, Shibata et al. (2008) reported that forest enclosures sometimes hamper tree regeneration within the fenced areas because of serious seed predation by increased mouse populations.

5. Conclusion

The difficulties faced in minimizing the effect of plant protection methods on ecosystems that have complex community interactions are shown in this chapter. To establish best practices for biodiversity conservation, adaptive management should be adopted. Within such frameworks, we should examine and revise protective management practices on the basis of scientific data assimilated from regular monitoring of such ecosystems, while also preferentially using metal-free plant protection materials.

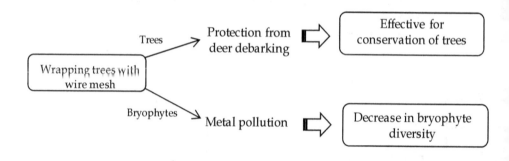

Figure 10. Positive and negative effects of wire mesh protection on the biodiversity

Author details

Yoshitaka Oishi
Department of Forest Science,
Faculty of Agriculture,
Shinshu University,
Japan

Acknowledgement

The author thanks Professor Yukihiro Morimoto and Associate Professor Hiroyuki Akiyama for providing critical comments and suggestions for improvements to this chapter, in addition to Dr. Kosaku Yamada for assistance with bryophyte identification and Kaori Kuriyama for the bryophyte survey and photos. This research was supported by a Grant-in-Aid for Scientific Research (A) (No. 18201008) from the Japan Society for the Promotion of Science and the Global COE Program "Global Center for Education and Research on Human Security Engineering for Asian Megacities," MEXT, Japan.

Appendix

This list shows the average cover of each species on a single *P. jezoensis* var. *hondoensis* tree. The cover levels are indicated for reference to the text. The bryophytes nomenclature follows that reported by Iwatsuki (2001).

Plots	A	B	C	D	E	F	G	H	I
Mosses									
Pogonatum alpinum (Hedw.) Röhl.	0	0	1.0	0	0.1	0.5	0	0	0.4
Pogonatum contortum (Brid.) Lesq.	0	0	0	0	0.1	0.1	0	0	0
Pogonatum japonicum Sull. & Lesq.	0	0	0	0	0.2	0	0	0	0
Dicranum japonicum Mitt.	0	0	0	0	0	0.3	0	0	0
Dicranum nipponense Besch.	0	0	0	0	0	0.2	0	0	0.2
Dicranum hamulosum Mitt.	0.3	0.1	0.5	0	1.2	0.7	0.1	0	0.6
Dicranum leiodontum Card.	1.0	0.6	1.0	0.3	0.8	0.6	0	0	0.6
Dicranum mayrii Broth.	0	0	0	0	0	0.1	0	0	0
Dicranum scoparium Hedw.	0	0.1	0	0	0.8	0.5	0	0	0.2
Dicranum viride (Sull. & Lesq.) Lindb. var. *hakkodense* (Card.) Takaki	1.0	0.4	1	0.5	0.9	0.9	0.4	0	0.4
Dicranodontium denudatum (Brid.) E. G. Britt. ex Williams	0	0.1	0.5	0	0.5	0.4	0	0	0.3
Dicranoloma cylindrothecium (Mitt.) Sakurai	0	0	0	0	0.3	1.2	0	0	0.8
Leucobryum bowringii Mitt.	0	0	0	0	0	0	0	0	0
Leucobryum juniperoideum (Brid.) Müll.Hal.	0	0	0	0	0.1	0.1	0	0	0
Trachycystis flagellaris (Sull. & Lesq.) Lindb.	0	0	0	0	0	0.3	0	0	0
Pterobryon arbuscula Mitt.	0	0	0	0	0	0.1	0	0	0
Neckera konoi Broth.	0	0	0	0	0	0	0	0	0
Fauriella tenuis (Mitt.) Card.	0	0	0	0	0.1	0.1	0	0	0
Thuidium tamariscinum (Hedw.) Shimp.	0	0	0	0	0	0	0	0	0
Plagiothecium euryphyllum (Card. & Thér.) Z.Iwats.	0	0	0.5	0	0	0.2	0	0	0
Heterophyllium affine (Hook.) M.Fleisch.	1.3	0.1	0.5	0	3.3	3.4	0.3	0	1.5
Brotherella fauriei (Card.) Broth.	0	0	0	0	0.1	0.3	0	0	0.2
Brotherella henonii (Duby) M.Fleisch.	0	0	0	0	0.1	0.2	0	0	0
Herzogiella turfacea (Lindb.) Z.Iwats.	0	0	0	0	0	0.1	0	0	0
Hypnum tristo-viride (Broth.) Paris	1.3	0.4	1.0	0.4	3.8	3.5	0.5	0	2.7
Hypnum fujiyamae (Broth.) Paris	0.3	0.1	0	0	0.5	0.5	0	0	0.5
Pseudotaxiphyllum pohliaecarpum (Sull. & Lesq.) Z.Iwats.	0	0	0	0	0.1	0.1	0	0	0
Hylocomium splendens (Hedw.) Schimp.	0	0	0	0	0	0.7	0	0	0.1
Pleurozium schreberi (Brid.) Mitt.	0	0	0	0	0	0.4	0	0	0.1

Liverworts

Herbertus aduncus (Dicks.) Gray	0.7	0.1	0.5	0	0.4	0.4	0	0	0.4
Trichocolea tomentella (Ehrh.) Dumort.	0	0	0	0	0	0	0	0	0
Blepharostoma trichophyllum (L.) Dumort.	0	0	0	0	0.7	0.9	0	0	0.4
Lepidozia reptans (L.) Dumort.	0	0	0	0	0.1	0.5	0	0	0.4
Lepidozia subtransversa Steph.	0	0	0	0	0	0.1	0	0	0
Lepidozia vitrea Steph.	0.3	0	0	0	0.8	1.0	0.1	0	0.7
Bazzania bidentula (Steph.) Steph.	0	0	0	0	0	0.1	0	0	0.1
Bazzania denudata (Torr. ex Lindenb.) Trevis.	0.3	0.1	0.5	0.1	1.1	1.0	0.1	0	0.9
Bazzania yoshinagana (Steph.) S.Hatt.	0	0	0	0	0.3	1.2	0.1	0	0.8
Cephalozia leucantha Spruce	0	0	0	0	0.1	0.1	0	0	0
Cephalozia lunulifolia (Dumort.) Dumort.	0	0	0	0	0.1	0	0	0	0
Chephaloziella sp.	0	0	0	0.1	0	0.2	0	0	0
Nowellia curvifolia (Dicks.) Mitt.	0	0	0	0	0.1	0.2	0	0	0
Odontoschisma denudatum (Mart.) Dumort.	0	0.1	0.5	0.1	0.8	0.6	0.1	0	0.2
Odontoschisma grosseverrucosum Steph.	0	0	0	0	0	0.1	0	0	0
Jamesoniella autumnalis (DC.) Steph.	0	0	0	0	0.3	0	0	0	0
Jungermannia subulata A.Evans	0	0	0	0	0	0.1	0	0	0
Anastrophyllum michauxii (F.Weber) H. Buch	0	0	1.0	0	0.6	0.4	0.1	0	0.3
Lophozia longiflora (Nees) Schiffn.	0	0	0	0	0.3	0.6	0.1	0	0.2
Lophozia incisa (Schrad.) Dumort..	0	0	0	0	0	0.5	0.1	0	0.2
Mylia verrucosa Lindb.	0	0	0	0	0	0.3	0	0	0.4
Scapania ampliata Steph.	0.3	0.3	0	0	0.4	1.9	0.4	0	1.2
Scapania bolanderi Austin	0	0	0	0	0	0.1	0	0	0.1
Scapania ciliata Sande Lac.	0	0	0	0	0.1	0	0	0	0
Scapania hirosakiensis Steph. ex Müll. Frib.	0	0	0	0.1	0.3	0.3	0	0	0
Heteroscyphus planus (Mitt.) Schiffn.	0	0	0	0	0.1	0	0	0	0
Plagiochila gracilis Lindenb. & Gottsche	0	0.1	0.5	0.3	0.2	0.5	0.2	0	0.1
Plagiochila ovalifolia Mitt.	0	0	0	0	0	0.1	0	0	0
Plagiochila semidecurrens (Lehm. & Lindenb.) Lindenb.	0	0	0	0.1	0.4	0.9	0.2	0	0.1
Radula cavifolia Hampe ex Gottsche, Lindenb. & Nees	0	0	0	0	0	0.1	0	0	0
Radula brunnea Steph.	0	0	0	0	0	0.4	0	0	0
Ptilidium pulcherrimum (Weber) Vain.	0	0.1	0	0	0.1	0	0	0	0
Frullania tamarisci (L.) Dumort. subsp. obscura (Verd.) S.Hatt.	0	0	0	0	0.6	1.6	0.2	0	1
Cololejeunea macounii (Spruce ex Underw.) A.Evans	0	0	0	0	0	0.1	0	0	0
Drepanolejeunea angustifolia (Mitt.) Grolle	0	0	1	0.1	0	0.4	0	0	0.1
Drepanolejeunea ternatensis (Gottsche) Steph.	0	0	0	0	0.1	0	0	0	0
Drepanolejeunea teysmannii Steph.	0	0	0	0	0	0.2	0.1	0	0
Nipponolejeunea pilifera (Steph.) S.Hatt.	1.0	0.7	0	0.4	1.0	0.9	0.2	0	0.9
Nipponolejeunea subalpina (Horik.) S.Hatt.	0.3	0	0	0.1	0.1	0.1	0	0	0.1
Lejeunea ulicina (Taylor) Gottsche, Lindenb. & Nees	0.3	0	0.5	0.1	0	0.5	0.1	0	0

6. References

Akashi. N. & Nakashizuka, T. (1999). Effects of bark-stripping by Sika deer (*Cervus nippon*) on population dynamics of a mixed forest in Japan. *Forest Ecolgy and Management*, Vol.113, No.1, (January 1999), pp. 75–82, ISSN 0378-1127

Allombert, S.; Gaston, A. J. & Martin, J.-L. (2005). A natural experiment on the impact of overabundant deer on songbird populations. *Biological Conservation*, Vol.126, No.1, (November 2005), pp. 1–13, ISSN 0006-3207

Ando, M. & Goda, R. (2009). Sika deer (*Cervus nippon* Temminck) of Mount Ohdaigahara, In: *Ecology of sika deer and forest ecosystem of Mt. Ohdaigahara*, Shibata, E. & Hino, T. (Eds.), 46-52 ,Tokai University Press, ISBN 978-4-486-01830-8, Kanagawa, Japan

Boudreault, C.; Coxson, D. S.; Vincent, E.; Bergeron, Y. & Marsh, J. (2008). Variation in epiphytic lichen and bryophyte composition and diversity along a gradient of productivity in *Populus tremuloides* stands of northeastern British Columbia, Canada. *Ecoscience*, Vol.15, No.1, (March 2008), pp. 101–112, ISSN 1195-6860

Coxson, D. S. (1991). Nutrient release from epiphytic bryophytes in tropical montane rain forest (Guadeloupe). *Canadian Journal of Botany*, Vol.69, No.10, (October 1991), pp. 2122–2129, ISSN 1480-3305

Doei, H. (1988). Bryophytes of Mt. Ohdaigahara in the Kii Peninsula, Central Honshu, Japan, I. *Nanki Biology*, Vol.30, pp. 14–23

Feber, R. E.; Brereton, T. M.; Warren, M. S. & Oates, M. (2001). The impacts of deer on woodland butterflies: the good, the bad and the complex. *Forestry*, Vol.74, No.3, pp. 271–276, ISSN 1464-3626

Flowerdew, J. R. & Ellwood, S. A. (2001). Impacts of woodland deer on small mammal ecology. *Forestry*, Vol.74, No.3, pp. 277–287, ISSN 1464-3626

Fritz, Ö. (2009). Vertical distribution of epiphytic bryophytes and lichens emphasizes the importance of old beeches in conservation. *Biodiversity and Conservation*, Vol.18, No.2, (October 2008), pp. 289–304, ISSN 1572-9710

Fuller, R. J. & Gill, R. M. A. (2001). Ecological impacts of increasing numbers of deer in British woodland. *Forestry*, Vol.74, No.3, pp. 193–199, ISSN 1464-3626

Harris, T. M. (1946). Zinc poisoning of wild plants from wire netting. *New Phytologist*, Vol.45, No.1, (June 1946), pp. 50–55, ISSN 1469-8137

Hazell, P.; Kellner, O.; Rydin, H. & Gustafsson, L. (1998). Presence and abundance of four epiphytic bryophytes in relation to density of aspen (*Populus tremula*) and other stand characteristics. *Forest Ecology and Management*, Vol.107, No.1–3, (August 1998), pp. 147–158, ISSN 0378-1127

Ide, H. & Kameyama, A. (1972). Vegetation of Ohdaigahara. *Applied phytosociology*, Vol.1, pp. 1–48.

Iwatsuki, Z. (Ed). (2001). *Mosses and liverworts of Japan*, Heibonsha, ISBN 978-4582535075, Tokyo, Japan

Japan Society of Corrosion Engineering. (Ed.) (2000). *Corrosion handbook*, Maruzen, ISBN 978-4621046487, Tokyo, Japan

Kumar, S.; Takeda, A. & Shibata, E. (2006). Effects of 13-year fencing on browsing by sika deer on seedlings on Mt. Ohdaigahara, central Japan. *Journal of Forest Research*, Vol. 11, No.5, (April 2006), pp. 337–342, ISSN 1610-7403

Lyons, B.; Nadkarni, N. M. & North, M. P. (2000). Spatial distribution and succession of epiphytes on *Tsuga heterophylla* (western hemlock) in an old-growth Douglas-fir forest. *Canadian Journal of Botany* , Vol.78, No.7, (June 2000), pp. 957–968, ISSN 1480-3305

Ministry of the Environment-Kinki Regional Environment Office. (2009). *Nature restoration plan for Mt. Ohdaigahara: second period*. Ministry of the Environment- Kinki Regional Environment Office, Osaka, Japan

Miquelle, D. G. & Van Ballenberghe, V. (1989). Impact of bark stripping by moose on aspen spruce communities. *The Journal of Wildlife Management*, Vol.53, No.3, (July 1989), pp. 577–586, ISSN 1937-2817

Nadkarni, N. M. (1984). Epiphyte biomass and nutrient capital of a neotropical elfin forest. *Biotropica*, Vol.16, No.4, (December 1984), pp. 249–256, ISSN 1744-7429

Nakajima, N. (2007). Changes in distributions of wildlife in Japan, In: *Rebellion of wildlife and collapse of forest*, Forest and Environment Research Association, Japan (Ed.), 57-68, Shinrinbunka Association, ISBN 978-4021001284, Tokyo, Japan

Nakamura, T. (1997). Effect of bryophytes on survival of conifer seedlings in subalpine, pp.forests of central Japan. *Ecological Research* Vol.7, No.2, pp. 155–162, ISSN 1440-1703

Nara local meteorological observatory. (1997). *Climate of Nara Prefecture in the last 100 years*. National printing bureau, Tokyo, Japan

Oishi, Y. (2011). Protective management of trees against debarking by deer negatively impacts bryophyte diversity. *Biodiversity and Conservation*, Vol.20, No.11, (June 2011), pp. 2527-2536, ISSN 1572-9710

Ojala, E.; Mönkkönen, M. & Inkeröinen, J. (2000). Epiphytic bryophytes on European aspen Populus tremula in old-growth forests in northeastern Finland and in adjacent sites in Russia. *Canadian Journal of Botany*, Vol.78, No.4, (April 2000), pp. 529–536, ISSN 1480-3305

Pellerin, S.; Huot, J. & Côté, S. D. (2006). Long term effects of deer browsing and trampling on the vegetation of peatlands. *Biological Conservation*, Vol.128, No.3, (March 2006), pp. 316–326, ISSN 0006-3207

Pypker, T. G.; Unsworth, M. H. & Bond, B. J. (2006). The role of epiphytes in rainfall interception by forests in the Pacific Northwest II. Field measurements at the branch and canopy scale. *Canadian Journal of Forest Reserch*, Vol.36, No.4, (April 2006), pp. 819–832, ISSN 1208-6037

R Development Core Team. (2010). *R 2.11.0, A language and environment for statistical computing computer program*, R Development Core Team, ISBN 3-900051-07-0, Vienna, Austria

Rooney, T. P. (2001). Deer impacts on forest ecosystems: a North American perspective. *Forestry*, Vol.74, No.3, pp. 201–208, ISSN 1464-3626

Rooney, T. P. & Waller, D. M. (2003). Direct and indirect effects of white-tailed deer in forest ecosystems. *Forest Ecology and Management*, Vol.181, No.1-2, (August 2003), pp. 165–176, ISSN 0378-1127

Salmon, T. P.; Whisson, D. A. & Marsh, R. E. (2006). *Wildlife Pest Control around Gardens and Homes* (2nd edition), University of California, Agricultural and Natural Resources Publication, ISBN 978-0931876660, California

Schütz, M.; Risch, A. C.; Leuzinger, E.; Krüsi, B. O. & Achermann, G. (2003). Impact of herbivory by red deer (*Cervus elaphus* L.) on patterns and processes in subalpine grasslands in the Swiss National Park, *Forest Ecolgy and Management* Vol.181, No.1-2, (August 2003), pp. 177–188, ISSN 0378-1127

Seaward, M. R. D. (1974). Some observations on heavy metal toxicity and tolerance in lichens. *The Lichenologist*, Vol.6, No.2, pp. 158–164, ISSN 0024-2829

Shibata, E.; Saito, M. & Tanaka, M. (2008). Deer-proof fence prevents regeneration of Picea jezoensis var.hondoensis through seed predation by increased wood mouse populations. *Journal of Forest Research*, Vol.13, No.2, (April 2008), pp. 89–95, ISSN 1610-7403

Stewart, G. H. & Burrows, L. E. (1989). The impact of white-tailed deer *Odocoileus virginianus* on regeneration in the coastal forests of Stewart Island, New Zealand. *Biological Conservation*, Vol.49, No.4, pp. 275–293, ISSN 0006-3207

Stockton, S. A.; Allombert, S.; Gaston, A. J. & Martin, J.-L. (2005). A natural experiment on the effects of high deer densities on the native flora of coastal temperate rain forests. *Biological Conservatoin*, Vol.126, No.1, (November 2005), pp. 118–128, ISSN 0006-3207

Takatsuki, S. (2009). Effects of sika deer on vegetation in Japan: A review. *Biological Conservation*, Vol.142, No.9, (september 2009), pp. 1922-1929, ISSN 0006-3207

Thomas, S. C.; Liguori, D. A. & Halpern, B. C. (2001). Corticolous bryophytes in managed Douglas-fir forests: habitat differentiation and responses to thinning and fertilization. *Canadian Journal of Botany*, Vol.79, No.8, (August 2001), pp. 886–896, , ISSN 1480-3305

Tyler, G.; Balsberg Påhlsson, A.-M.; Bengtsson, G.; Bååth, E. & Tranvik, L. (1989). Heavy metal ecology of terrestrial plants, microorganisms and invertebrates. *Water, Air, and Soil Pollution*, Vol.47, No.3-4, pp. 189–215, ISSN 1573-2932

Tyler, G. (1990). Bryophytes and heavy metals: a literature review. *Botanical Journal of the Linnean Society*, Vol.104, No.1-3, (September 1990), pp. 231–253, ISSN 1095-8339

Vercauteren, K. C.; Michael, J. L. & Hygnstrom, S. (2006). Fences and deer-damage management: A review of designs and efficacy. *The Wildlife Society Bulletin*, Vol.34, (August 2006), pp. 191–200, ISSN 0091-7648

Ward, J. S.; Gent, M. P. N. & Stephens, G. R. (2000). Effects of planting stock quality and browse protection-type on height growth of northern red oak and eastern white pine. *Forest Ecology and Management*, Vol.127, No.1-3, (March 2000), pp. 205–216, ISSN 0378-1127

Webster, C. R.; Jenkins, M. A. & Rock, J. H. (2005). Long-term response of spring flora to chronic herbivory and deer exclusion in Great Smoky Mountains National Park, USA. *Biological Conservavion*, Vol.125, No.3, (October 2005), pp. 297–307, ISSN 0006-3207

Williams, C. B. & Sillett, S. C. (2007). Epiphyte communities on redwood (Sequoia sempervirens) in northwestern California. *The Bryologist*, Vol.110, No.3, pp. 420– 452, ISSN 0007-2745

Yokoyama, S.; Maeji, I.; Ueda, T.; Ando, M. & Shibata, E. (2001). Impact of bark stripping by sika deer, Cervus nippon, on subalpine coniferous forests in central Japan. *Forest Ecolgy and Management*, Vol.140, No.2-3, (January 2001), pp. 93–99, ISSN 0378-1127

Effect of *Pseudotsuga menziesii* Plantations on Vascular Plants Diversity in Northwest Patagonia, Argentina

I. A. Orellana and E. Raffaele

Additional information is available at the end of the chapter

1. Introduction

Forests biodiversity conservation is a global concern because they are home to 80% of the biodiversity of terrestrial environments [1, 2, 3, 4]. The replacement of all or part of these ecosystems with monocultures creates mosaics of vegetation, contributing to habitat fragmentation [5]. The new landscape includes more homogeneous vegetation units and may differ in patterns and processes from the original landscape formed by primary or secondary forests.

During the second half of the twentieth century there was an increase of areas subjected to forestations with exotic species. It is estimated that about 187 millions ha were planted worldwide, which represents 5% of global forests [2]. The annual growth of forest plantations worldwide is estimated at 2-3 million ha per year [6]. Sixty percent (60%) of forest plantations are located in four countries: China, India, Russian Federation and United States. In the southern hemisphere, emerging forestry countries are: Brazil, New Zealand, Chile, South Africa, Argentina, Uruguay, Venezuela and Australia [7]. In non-tropical areas, a third of the area of native forest destroyed is used for forest plantations [3].

In the West of Chubut, Río Negro and Neuquén provinces in Argentina, the forested area with exotic species reached 70,000 ha in 2007 [8]. The species used are *Pinus ponderosa*, *Pseudotsuga menziesii*, *Pinus radiata* and *Pinus contorta*. In the late 1990's it was estimated that forestation had a rates of 10,000 ha per year [9, 10]. Exotic plantations effects on biological diversity in *Austrocedrus chilensis* forests and mixed shrublands in Temperate Forests of South America are still poorly understood.

Several studies have shown that species diversity decreases in areas of forestations, and seems to depend on the proximity of plantations to native environments and treatments prior to afforest. In Congo, diversity of vascular plants was compared in mixed stands of *Eucalyptus - Acacia - Pinus*, with secondary forests and the African savannah which showed a reduction in the number of species in plantations understory at compared to the understory of secondary forest. At the plantations edges, however, the loss of species richness was lower, so that proximity to the pre-existing forest help to maintain diversity over short distances [11]. In a conservation rainforest study, the richness of species was found to be lower in *Coffea arabica* and *Elettaria cardamomum* plantations than in the native forest [12]. On the other hand, at landscape scale, there has been that *E. cardamomum* plantations where shrubs and herbs strata were retained, connectivity was maintained among fragmented forest patches. Similar results were found when analyzing the feasibility of employing commercial pine plantations as complementary habitat to conserve threatened species in Chile [13, 14].

Besides the proximity to native environments, the diversity in forestations it was found related to age. For example, in 20 years old *P. radiata* plantations in New Zealand, *Rubus fruticosus* and other generalist species were frequently found while in later succession (40 years), *R. fruticosus* is replaced by several species of shade-tolerant native ferns and shrubs [15]. Similar results were found in other older plantations of *P. radiata* in New Zealand, where diversity of vascular plants was similar in plantations and nearby native forests, which also was confirmed that plantations provided habitat for some species of birds such as *Apteryx mantelli* [16, 17]. In some cases, older plantations increase the supply of habitat, increases in spatial and vertical heterogeneity, increases in light levels, development of organic soil layers and the associated fungal flora [4]. However, there are some studies that show that diversity does not increase with age of forest plantations. The diversity of the beetles ensemble was lower in older plantations of *Picea abies* than in younger plantations [18].

In Chilean temperate forests, where the replacement of native forests by exotic forestations was important, there was a decrease in the distribution of endangered and vulnerable vascular plants [19, 20]. There was also a loss of *Nothofagus* native forest structure, with the disappearance of strata, as well as decreases in species richness of vascular plants [21, 22]. Similar patterns with decreases in vascular plants, beetles and birds species were found in mixed plantations of *P. menziesii*, *P. radiata* and *Pinus sylvestris* installed in *Nothofagus* forests and *P. ponderosa* plantations instaled in the steppe in Argentina [23, 24]. In other studies it was found that the richness and composition of birds was more affected by the structure of plantations than by their tree species composition [25, 26]. While a study on ant assemblages showed that in plantations there are decreases in abundance and changes in the composition of species respect to the nearby steppe [27]. All these studies support the hypothesis that high individual density in forest plantations affect biodiversity within them, and propose lower density of trees as an alternative to improve biodiversity.

Diversity loss in South America temperate forests, is a topic of great interest since these ecosystems are characterized by high levels of endemism, a product of a deep biogeographic isolation with common ancestry biota, as well as its extremely heterogeneous floristic composition, derived from various biogeographical sources (e.g. Gondwana, Neotropical, Boreal) [28, 29]. The vascular flora has about 34% of woody genera endemic. Most of the endemisms are monotypic with only one species per genus [28]. *Austrocedrus chilensis* is an endemic monotypic species of South America Temperate Forests, with a smaller distribution area than that occupied in ancient geological times [30, 31]. At the present, this species has serious conservation problems due to multiple anthropogenic disturbances, and is included in the IUCN Red List in the "vulnerable" status [32]. This chapter presents some results related to vascular plants diversity in *A. chilensis* forests, and mixed shrublands when they are replaced by the exotic conifer *P. menziesii* plantations.

1.1. Hypothesis

In *P. menziesii* forestations in Patagonia there is reduced vascular plant diversity compared with the natural communities they replace.

1.2. General aim

Study and compare the vascular plants diversity in *P. menziesii* forestations and *A. chilensis* forests and mixed shrublands adjacent in the Northwest of Chubut Province and Southwest of the Río Negro Province, Argentina.

1.3. Specific aims

1.   To estimate the alpha and beta diversity of vascular plants in *P. menziesii* plantations and contiguous *A. chilensis* forests and mixed shrubland.
2.   To analyze the similarity in the composition and abundance of herbaceous and woody shrubs growing in *P. menziesii* forestations, *A. chilensis* forests and mixed shrublands.

2. Methods

2.1. Study system

The study area includes the West of Chubut and the Southwest of the Río Negro Provinces, in Argentina, between the localities of Corcovado 43º 32' 36.54" South, 71° 26' 37.5" West and San Carlos de Bariloche 41º 8' 16.83 " South, 71° 17' 12.09" W (Fig. 1). In this area there are about 103 *P. menziesii* plantations. The planted surfaces vary between 0.5 and 12 ha, and initial densities are 1.000 trees per ha [9, 33]. The age of the plantations of *P. menziesii* studied for 2006, ranged between 17 and 35 years old, and all had reached reproductive maturity. *P. menziesii* is native to North America where it is distributed between 55° and 19º N, in temperate climates [34]. In Patagonia, Argentina, *P. menziesii* plantations were installed in a range of precipitations between 1500 mm to 600 mm. In this area, various native plant communities were replaced by afforestations, but the best growths are associated with the natural range of the forests of *A. chilensis* and the mixed shrublands, so that these environments have a higher substitution pressure.

The mixed shrublands are characterized by a shrub stratum of 5 to 7 m high, in which the most abundant species are: *Diostea juncea, Lomatia hirsuta, Embothrium coccineum, Schinus patagonicus, Fabiana imbricata* and some isolated trees of *A. chilensis* and *Maytenus boaria*. It is also distinguished is a shrub sub-stratum with similar species composition, and a herbaceous stratum, dominated by species of the families *Poaceae, Asteraceae* and *Rosaceae*. In *A. chilensis* forests a distinguished tree stratum of 15 m in height is found, in which *A. chilensis* is the dominant species. The shrub stratum was mainly composed by *S. patagonicus, L. hirsuta*, and *E. coccineum*, among others. *Asteraceae* and *Poaceae* dominate the herbaceous stratum.

2.2. Sampling design

Four sites were selected where *P. menziesii* plantations were adjacent to mixed shrublands and eleven sites of *P. menziesii* plantations adjacent to *A. chilensis* forests (Fig. 1).

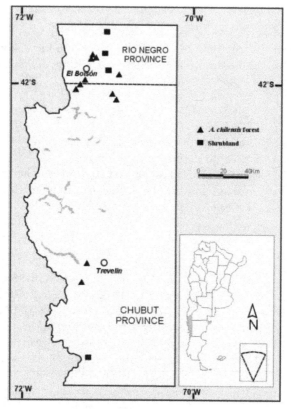

Figure 1. Map of the study area and locations of study sites.

On each site (plantation-native community edge area) a transect perpendicular to the edge line was established (Fig. 2). Each transect was subdivided into 11 plots of 100 m², three plots were installed in plantations at -30, -20 and -10 m from the edge line, and eight in the native communities at 10, 20 , 30, 40, 50, 60, 70 and 80 m from the edge line.

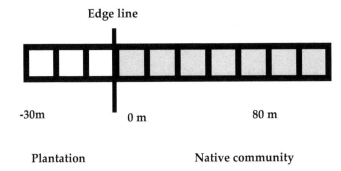

Figure 2. Scheme of a sampling transect established in a plantation-native community edge area, sampling units (plots) and edge line are indicated.

In each 100 m² plots, the composition and abundance of vascular plants was registered and mesured, according to the following classification:

1. herbaceous stratum: included herbaceous and woody species seedlings below 10 cm height. The measurements were performed using four 0.5 m² circular plots randomly selected within each sampling unit of 100 m². The species cover percentage was measured, and individual numbers for woody seedlings counted.
2. shrub stratum: included shrubs, woody vines and woody species saplings above 10 cm in height and below 5 cm in diameter at breast height (DBH). The number of individuals by species was counted on a plot of 25 m² randomly selected in each 100 m² plot.
3. tree stratum: includes trees and shrubs greater than 5 cm DBH. All individuals were counted in the 100 m² plot.

Vascular plant field samples were collected and identified in the laboratory. The reference of support was Patagonian Flora collection volume VIII, Parts I, II, III, IVa and b, V, VI and VII [35].

2.3. Alpha diversity data analysis

The Alpha diversity analysis was made by species accumulation curves, by using EstimateS software [36, 37]. The vascular plants diversity was analyzed in *P. menziesii* plantations and in the adjacent mixed shrublands, as in plantations and *A. chilensis* forests contiguous. The Clench equation was used, which has demonstrated good fits for multiple sampling of taxa of species [$S_n = a*n/ (1+b*n)$], where:

S_n = mean number species for each sampling unit
a = increase rate of new species at the beginning of samplings
b = is a parameter related to the shape of the curve
n = number of sampling units

This function was applied using a nonlinear estimation, through the iterative adjustment of Simplex and Quasi-Newton algorithm with Statistica 7 [38, 39]. The fit of the equation to the accumulation curve, was analyzed by calculating the Determination Ccoefficient: R^2. The slope of the curve (S) when the number of samples is maximum was used to assess sampling quality $[S = a / (1 + bn)^2]$ [39]. The flora proportions recorded provided further information on vegetation sampling quality $[[S_{obs.}/(a/b)] * 100]$ [39].

2.4. Beta diversity data analysis

The diversity between habitats was analyzed by using the Jaccard similarity index [37]. This index calculated the species replacement degree across environmental gradients. Where environments are very different, and there are no shared species a 0 value occurs. If all species are shared the index value is set to 1.

This study the Jaccard similarity index $[I_J = c/ a+b-c]$ was obtained for the following pairs of communities: *P. menziesii* plantations (A)-mixed shrublands (B), *P. menziesii* plantations (A) – *A. chilensis* forests (B).

Where:

a = vascular plant species number in A community
b = vascular plants species number in B community
c = vascular plants species number in both A and B communities.

2.5. Analysis of similarity in composition and abundance of herbaceous and shrubs species strata by ANOSIM and MDS

In order to determine if plantations affected the species composition of native communities, the similarity in composition and abundance of herbaceous and shrubs in plantations and native communities were analized by using the multivariate ANOSIM method [40]. This analysis performs permutations on similarity matrices and produces a statistic (R) which is an absolute measure of distance between groups. An R value close to 1 indicates that the assemblages are very different, while a R value close to 0 indicates that the assemblages are similar [40]. To illustrate the assembly of herbaceous and shrub species found in plantations and native communities the Non-Metric Multidimensional Scaling (NMDS) method was applied with the similarity index of Bray-Curtis. These analyzes were performed using the statistical program PRIMER-E Ltd. [41].

3. Results

3.1. Alpha diversity analysis in plantations and native communities

The species accumulation curves obtained from mixed shrublands and *P. menziesii* plantations are presented in Figure 3. Both curves show how species richness increases with increasing number of sampling units. They also show that species richness is higher in mixed shrublands that in contiguous *P. menziesii* plantations.

Model **(1)** describes the species accumulation curve in the mixed shrublands, the model **(2)** describes species accumulation curve in adjacent *P. menziesii* plantations (n = number of sampling units).

$$S_{n \text{ mixed shrublands}} = 18,4 * n / (1 + 0,2 * n) \quad R^2 = 0,99 \tag{1}$$

$$S_{n \text{ P. menziesii plantations}} = 3,3 * n / (1 + 0,22 * n) \quad R^2 = 0,99 \tag{2}$$

The Determination Coefficients (R^2) of both models indicate that adjustments to the models accumulation curves are highly representative. Those models were used to calculate the slope of the tangent line when it reaches the maximum number of sampling units. For the species accumulation curve obtained in *P. menziesii* plantations, the slope was **S** = 0.25 (n = 12), so it would have been possible to add new species by increasing the sampling units number. The proportion of vascular plants recorded in *P. menziesii* plantations was 73%. In mixed shrublands, the slope was **S** = 0.33 (n = 32) with a proportion of 86% of vascular plants recorded.

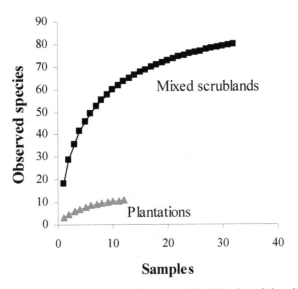

Figure 3. Species accumulation curves obtained from mixed shrublands and the adjacent *P. menziesii* plantations.

The species accumulation curves obtained from *A. chilensis* forests and *P. menziesii* plantations also show a greater species richness in *A. chilensis* forests that in *P. menziesii* plantations (Fig. 4).

Model **(3)** describes the species accumulation curve in *A. chilensis* forests, model **(4)** describes the species accumulation curve in adjacent *P. menziesii* plantations (n = number of sampling units).

$$S_{n \text{ A. chilensis forests}} = 13,05 * n / (1 + 0,07 * n) \quad R^2 = 0,99 \tag{3}$$

$$S_{n \text{ P. menziesii plantations}} = 5,71 * n / (1 + 0,05 * n) R^2 = 0,99 \tag{4}$$

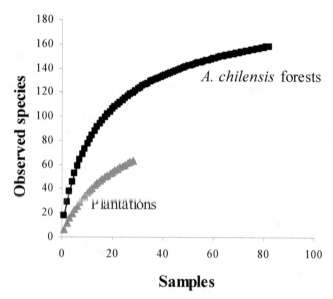

Figure 4. Species accumulation curves obtained from *A. chilensis* forests and the adjacent *P. menziesii* plantations.

The Determination Coefficients (R^2) of both models indicate that adjustments to the models accumulation curves are highly representative. For the species accumulation curve obtained in *P. menziesii* plantations, the slope was **S** = 0.88 (n = 28), so it would have been possible to add new species by increasing the number of sampling units. The sampling registered only 60% of flora of *P. menziesii* plantations (n = 28). In the *A. chilensis* forests contiguous afforestations, for n = 82, **S** = 0.28, so that also have been possible to obtain a greater number of species by increasing the number of sampling units. There, sampling showed 85% of flora of *A. chilensis* forests.

3.2. Beta diversity

When similarity was analyzed between *P. menziesii* plantations and mixed shrublands, the Jaccard index was 0.14 ± 0.02 (± SE), indicating that plantations and mixed shrublands contiguous are dissimilar in species composition. In the same way *P. menziesii* plantations and *A. chilensis* forests showed that both adjoining communities were dissimilar in vascular plants composition, the Jaccard index reached a value of 0.17 ± 0.04 (± SE).

3.3. Floristic similarity between *P. menziesii* plantations, *A. chilensis* forests and mixed shrublands

According to the R value obtained, herbaceous species composition was similar in the three communities studied (R = -0.106, p = 4.3%). However, the confidence level for the analysis was not significant (p <0.1%). The greatest similarities in herbaceous species composition were recorded from *A. chilensis* forests and adjoining plantations, while communities with greater differences in herbaceous species composition are plantations and mixed shrublands (Table 1).

Groups	R	Significance levels (%)
F-S	0,256	95,6
F-P	-0,02	65,4
S-P	-0,366	**99,6**

Table 1. Analysis of similarity (ANOSIM) of herbaceous species found in *P. menziesii* plantations (P), *A. chilensis* forests (F), and mixed shrublands (S). R value is a measure of similarity of species in different communities, R values close to 1 indicate differences, and 0 indicate similarities in species composition. Significant values highlighted in bold (p < 0.1%).

Non-Metric Multidimensional Scaling (NMDS) analysis, showed greater variability in the composition and abundance of herbaceous species in the differente sites corresponding to *P. menziesii* plantations. Most points (sites) were located towards the periphery of the graph and away from each other (Fig. 5). While in the center of the graph the cloud of points corresponding to mixed shrublands and *A. chilensis* forest sites. showed a more similar species composition between both native plant communities. Stress = 0.13, equivalent to a good level of confidence.

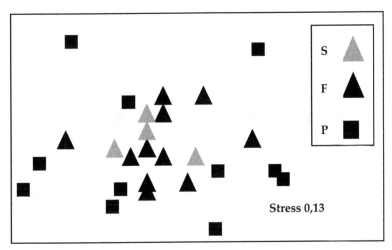

Figure 5. Graphical representation of herbaceous composition and abundance (NMDS) in different communities: F = *A. chilensis* forests, P = *P. menziesii* plantations, and S = mixed shrublands.

Shrub species composition was similar between the communities (R = 0.2, p = 0.4%). The more similar communities to each other on shrub species composition were mixed shrublands and *A. chilensis* forests, but as the above analysis, the confidence level was not significant. The least similar communities were forests and plantations (Table 2).

The NMDS analysis shows two clouds of points, one to the right and the other one to the left of the graph (Fig. 6).The right cloud includes points corresponding to *P. menziesii* plantations. This pattern indicates that composition and abundance of shrub species assemblages is more similar between plantations than between plantations and neighboring native communities. To the left, the points cloud is less scattered, includes sites of *A. chilensis* forests and mixed shrublands. This analysis shows greater similarity in the assembly of vascular plants in *A. chilensis* forests and mixed shrublands.

Groups	R	Significance levels (%)
F-S	-0,13	78,6
F-P	0,34	0,1
S-P	0,20	10,2

Table 2. Analysis of similarity (ANOSIM) of shrubs species found in *P. menziesii* plantations (P), *A. chilensis* forests (F), and mixed shrublands (S). R value indicates the species similarity between the different communities, R values close to 1 indicate differences, and 0 indicate similarities in species composition.

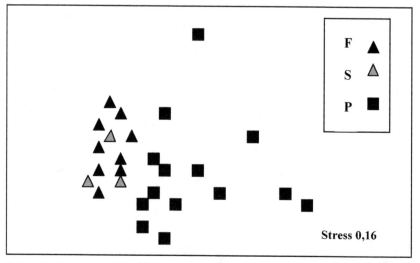

Figure 6. Graphical representation of the ordination (NMDS) of shrub stratum species in different communities: F = *A. chilensis* forests, P = *P. menziesii* plantations, and S = mixed shrublands.

4. Discussion and conclusions

In the plant communities studied, the greatest vascular plant diversity was found in *A. chilensis* forests, where 168 vascular plants species were recorded, while in adjacent *P. menziesii* plantations were 37.5% of this number of species. In mixed shrublands 86 vascular plants species were recorded, and only 13% of this species number in contiguous *P. menziesii* plantations. Species accumulation curves allowed comparison of species richness in adjacent communities: *P. menziesii* plantations- *A. chilensis* forests, and *P. menziesii* plantations- mixed shrublands. Greater species richness was observed in native communities than in plantations, confirming that there is a loss of vascular plant diversity in *P. menziesii* plantations.

While the similarity analysis (ANOSIM) showed a similar species composition between all pairs of communities, these analyses was not significant. The similarity analysis based on the composition and abundance (NMDS) for herbaceous and shrub strata, showed that both mixed shrublands and *A. chilensis* forests are more similar to each other than with the contiguous *P. menziesii* plantations. It is striking that different native communities (forests and shrublands) separated by distances of up to 400 km presented the highest similarity of vascular plants. If there was no effect of plantations on vascular plants composition and abundance, a greater similarity between native community and their corresponding neighboring plantation would be expected. In general terms, this analysis confirms that *P. menziesii* plantations changed the native communities with a noticeable loss of diversity and changes in the abundance and composition of vascular plant.

The vascular plants loss in plantations adjacent to mixed shrublands can be explained by decreases in radiation and environmental heterogeneity in plantations [42]. The environmental heterogeneity is an important factor that promotes biodiversity [23, 43]. Differences in radiation between shrublands and plantations, are stronger than differences between *A. chilensis* forests and *P. menziesii* plantations. So understory species of *A. chilensis* forests, could find a more similar environment in adjacent plantations, than species in mixed shrublands [42]. Furthermore, *A. chilensis* forests are a major source of vascular plants diversity, which may spread to adjoining plantations. While the native communities proximity to plantations promotes species dispersal towards plantations, environmental conditions in plantations prevent the establishment.

The diversity loss found in *P. menziesii* plantations that replaced *A. chilensis* forests and mixed shrublands is similar to the results found in Chile, South Africa, New Zealand and Argentina [11, 19-23, 44, 45]. All these studies support the idea that there is a loss of diversity in forest plantations. This study shows the effects on vascular plants diversity when native ecosystems are replaced by exotic species forestations. In addition to changes in vascular plant diversity in forested areas, there are other edge effects that alter vascular plant structure in the native communities surrounding *P. menziesii* plantations [42]. One of the most important processes recorded in the edge areas is the establishment and dispersion of *P. menziesii* seedlings and saplings from *P. menziesii* plantations [46, 47]. These invasion processes together with diversity loss contribute to native ecosystems degradation.

Author details

I. A. Orellana
Andean Patagonian Forest Research and Extension Center (CIEFAP),
Patagonia San Juan Bosco National University (UNPSJB), Esquel, Chubut, Argentina

E. Raffaele
Ecotone Laboratory, Comahue National University - INIBIOMA/ CONICET,
S. C. de Bariloche, Río Negro, Argentina

Acknowledgement

This manuscript was supported by Universidad Nacional de la Patagonia San Juan Bosco PI N° 560, Centro de Investigación y Extensión Forestal Andino Patagónico (CIEFAP), and PICTO Forestal N°36879 MinCyT Argentina. We would like to thank J. Monges for his field assistance and Javier Puig, Guillermo Defosse and Mark Austin for their help with the language.

5. References

[1]   Naciones Unidas (1992) Convention on Biological Diversity. New York; Environmental Policy and Law.

[2]   FAO (2001) State of the World's forests, 2001. Roma, Italia: Food and Agriculture Organization of the United Nations.

[3]   Carnus JM, Parrotta J, Brockerhoff EG, Arbez M, Jactel H, Kremer A Lamb D, O'Hara K, Walters B (2003) Planted Forests and Biodiversit. UNFF Intersessional Experts Meeting on the Role of Planted Forests in Sustainable Forest Management, New Zealand.

[4]   Brockerhoff EG, Jactel H, Parrotta JA, Quine CP, Sayer J (2008) Plantation forests and biodiversity: oxymoron or opportunity? Biodiversity Conservation, 17, 925–951.

[5]   SchelhasJ, Greenberg R (1996) The value of forest patches. In: J. Schelhas y R. Greenberg (Eds.), Forest patches in tropical landscapes (pp. xv – xxxvi). Washington DC: Island Press.

[6]   FAO (2006). Global Forest Resources Assessment 2005—Progress towards sustainable forest management. FAO Forestry Paper 147. Roma, Italia: Food and Agriculture Organization of the United Nations.

[7]   Sedjo RA, Lyon KS (1990) The long term adequancy of wordl timber supply. Resource for the future.Washington DC

[8]   Rusch V, Vila A (2007) La conservación de la biodiversidad en ambientes bajo uso productivo. I reunión sobre forestación en la Patagonia. Esquel, Argentina: CIEFAP.

[9]   Davel M (2008) Estimación de la productividad de sitio. In: M. Davel (Ed.), Establecimiento y manejo del pino oregón en Patagonia (pp. 30-42). Esquel, Argentina: CIEFAP.

[10] Schlichter T, Laclau P (1998) Ecotono estepa-bosque y plantaciones forestales en la Patagonia norte. Ecología Austral, 8, 285-296.

[11] Huttel C, Loumeto JJ (2001) Effect of exotic tree plantations and site management on plant diversity. In: F. Bernhard-Reversat (Ed.), Effect of Exotic Tree Plantations on Plant Diversity and Biological Soil Fertility in the Congo Savanna: With Special Reference to Eucalypts, (pp. 9-18). Bogor, Indonesia: Center for International Forestry Research.

[12] Raman TRS (2006) Efects of habitat structure and adjacent habitats on birds in tropical rainforest fragments and shaded plantations in the Western Ghats, India. Biodiversity Conservation, 15, 1577–1607.

[13] Saavedra B, Simonetti JA (2005) Small mammals of Maulino forest remnants, a vanishing ecosystem of south-central Chile. Mammalia, 69 (3-4), 337-348.

[14] Vergara PM, Simonetti JA (2006) Abundance and movement of understory birds in a maulino forest fragmented by pine plantations. Biodiversity and Conservation, 15 (12), 3937-3947.

[15] Brockerhoff EG, Ecroyd CE, Leckie AC, Kimberley MO (2003) Diversity and succession of adventive and indigenous vascular understorey plants in *Pinus radiata* plantation forests in New Zealand. Forest Ecology and Management, 185, 307–326.

[16] Allen .B, Platt KH, Coker REJ (1995) Understorey species composition patterns in a *Pinus radiata* D. Don plantation on the central North Island volcanic plateau, New Zealand. New Zealand Journal of Forestry Science, 25, 301-317.

[17] Kleinpaste R (1990) Kiwis in a pine forest habitat. In: E. Fuller (Ed.), Kiwis, a Monograph of the Family Apterygidae (pp. 97-138). Auckland, New Zealand: Seto Publishing.

[18] Magura T, Tothmeresz B, Elek Z (2006) Changes in carabid beetle assemblages as Norway spruce plantations age. Community Ecology, 7(1), 1-12.

[19] Benoit I (1989) Libro Rojo de la Flora Terrestre de Chile. Corporación Nacional Forestal, Ministerio de Agricultura, Santiago de Chile

[20] Lara A, Donoso PJ, Cortés M (1991) Development of conservation and management alternatives for native forest in south-central Chile. Santiago, Chile: Reporte Final WWF-US-CODEFF.

[21] Von Buch MW, Osorio M (1987) Probleme um die Pinus radiata-Monokulturen in Südchile. Forstarchiv, 58, 249-253.

[22] Frank D, Finckh M (1997) Impactos de las plantaciones de pino oregón sobre la vegetación y el suelo en la zona centro-sur de Chile. Revista Chilena de Historia Natural, 70, 91-211.

[23] Raffaele, E. y Schlichter, T. (2000). Efectos de las plantaciones de pino ponderosa sobre la heterogeneidad de micrositios en estepas del noroeste patagónico. Ecología Austral, 10, 151-158.

[24] Paritsis J, Aizen MA (2008) Effects of exotic conifer plantations on the biodiversity of understory plants, epigeal beetles and birds in Nothofagus dombeyi forests. Forest Ecology and Management, 255, 1575–1583.

[25] Lantschner MV, Rusch V (2007) Impacto de diferentes disturbios antrópicos sobre las comunidades de aves de bosques y matorrales de *Nothofagus antarctica* en el NO Patagónico. Ecología Austral 17, 99-112.

[26] Lantschner MV, Rusch V, Peyrou C (2008) Bird Assemblages in Pine Plantations Replacing Native Ecosystems of N.W. Patagonia, Argentina. Biodiversity and Conservation 17(5), 969-989.

[27] Corley JC, Sackmann P, Rusch V, Bettinelli J, Paritsis J (2006) Effects of pine silviculture on the ant assemblages (Hymenoptera: Formicidae) of the Patagonian steppe. Forest Ecology and Management 222, 162-166.

[28] Arroyo MK, Cavieres L, Peñaloza A, Riveros M, Faggi AM (1996) Relaciones fitogeográficas y patrones regionales de riqueza de especies en el bosque templado lluvioso de Sudamérica. In: J. Armesto, C. Villagrán y M.K. Arroyo (Eds.), Ecología de los bosques nativos de Chile (pp. 71-92). Santiago, Chile: Editorial Universitaria.

[29] Armesto JJ, Lobos PL, Arroyo MK (1996) Los bosques templados del sur de Chile y Argentina: Una Isla Biogeográfica. In: J. Armesto, C. Villagrán y M.K. Arroyo (Eds.), Ecología de los bosques nativos de Chile (pp. 23-27). Santiago, Chile: Editorial Universitaria.

[30] Cabrera AL (1971) Fitogeografía Argentina. Boletín de la Sociedad Argentina de Botánica, 14(1), 36-38.

[31] Donoso C, Escobar B, Pastorino M, Gallo L, Aguayo J (2006) *Austrocedrus chilensis* (D. Don) Pic. Ser. Et Bizzarri Ciprés de la Cordillera, Len. In: C. Donoso (Ed.), Las especies arbóreas de los bosques templados de Chile y Argentina: Autoecología (pp.54-67). Valdivia, Chile: Mariza Cuneo Ediciones.

[32] IUCN Red List of Threatened Species (2012). Available: www.iucnredlist.org

[33] Gonda H, Mohr Bell D, Sbrancia R, Lencinas J, Bava J, Monte C, Mortero A, Sieber A (2009) Inventario del bosque implantado de la Provincia del Neuquén. Argentina. Patagonia Forestal, 14 (4), 13-16.

[34] Hermann RK, Lavender DP (1990) *Pseudotsuga menziesii* (Mirb.) Franco. In: R.M. Burns y B.H. Honkala (Eds.) Silvics of Noth America Volume 1 Conifers. Agriculture Handbook 654 (pp. 527-540). Washington, USA: Forest Service.

[35] Correa MN (1971 1999) Flora Patagónica VIII (I, II, III, IVa, IVb, V, VI, VII). Buenos Aires, Argentina: INTA.

[36] Gotelli NJ, Colwell RK (2001) Quantifying biodiversity: procedures and pitfalls in the measurement and comparison of species richness. Ecology Letters, 4, 379-391.

[37] Moreno CE (2001) Métodos para medir la biodiversidad. (Manuales y Tesis 1). Zaragoza, España: Sociedad Entomológica Aragonesa.

[38] Soberón J, Llorente J (1993) The use of species accumulation functions for the prediction of species richness. Conservation Biology, 7, 480-488.

[39] Jiménez Valverde A, Hortal J (2004) Las curvas de acumulación de especies y la necesidad de evaluar la calidad de los inventarios biológicos. Revista Ibérica de Aracnología, 8(31), 151-161.

[40] Clarke KR, Gorley RN (2001) PRIMER V5: User manual tutorial. Plymouth, United Kingdom: Plymouth Marine Laboratory.

[41] Clarke KR (1993) Non-parametric multivariate analyses of changes in community structure. Australian Journal of Ecology, 18, 117-143.

[42] Orellana I (2011) Efecto de borde de las plantaciones de *Pseudotsuga menziesii* sobre comunidades vegetales naturales en el noroeste patagónico. PhD thesis Universidad Nacional del Comahue, Argentina.

[43] Rosenzweig ML, Abramsky Z (1993) How are diversity and productivity related?. In: R.E. Ricklefs y D. Schuler (Eds.), Species diversity in ecological communities (pp. 52-65). Chicago: University of Chicago Press.

[44] Ogden J, Braggins J, Stretton K, Anderson S (1997) Plant species richness under *Pinus radiata* stands on the central north island volcanic plateau New Zealand. New Zealand Journal of Ecology, 21(1), 17-29.

[45] Barbaro L, Rossi JP, Vetillard F, Nezan J,Jactel H (2007) The spatial distribution of birds and carabid beetles in pine plantation forests: the role of landscape composition and structure. Journal of Biogeography 34:652–664.

[46] Sarasola MM, Rusch VE, Schlichter TM, Ghersa CM (2006) Invasión de coníferas forestales en áreas de estepa y bosques de ciprés de la cordillera en la Región Andino Patagónica. Ecología Austral, 16(2), 143-156.

[47] Orellana IA, Raffaele E (2010). The spread of *Pseudotsuga menziesii* in the *Austrocedrus chilensis* forest and shrubland, in the North Western Patagonia. New Zealand Journal of Forest Science, 40: 199-209.

Health Related Matter

Chapter 9

Biodiversity and Mental Health

Hector Duarte Tagles and Alvaro J. Idrovo

Additional information is available at the end of the chapter

1. Introduction

Humans depend on goods and services provided by natural environments for a decent, healthy, and secure life [1]. There is an increasing evidence of the health benefits to the people exposed to natural environments [2]. Physical health improvement by exposure to natural environments has been attributed mainly to the access and motivation of people to engage in physical activities (the so-called "green exercise" [3]), although, some controversies still remain [4]. It is well-known the positive association between physical activity and health by improving the physical fitness of people [5], and some studies have reported the beneficial impact of exercising to mental health as well [6].

Other studies have found the association between improved mental health and natural environment exposure by psychological mechanisms of restoration, rather than through mere physical exertion [7], or by enhancing social cohesion [8]. However, most of the studies investigating the association between natural environment exposure and mental health had focused on urban settings (green areas) of developed countries, where social, demographic, and geographic contexts, may be different from those of less developed economies. There is a lack of studies about the association between exposure to natural environments or green spaces and mental health in medium to low-income countries. A report of a cross-national prevalence of major depressive episodes, showed a significant higher lifetime prevalence in high-income countries than medium to low-income countries [9]. However, no significant difference in 12-month prevalence of major depressive episode was found.

The chapter begins with definitions about biodiversity and provides some arguments of concern for its current status. Then, from a theoretical and empirical perspective, it is explained the general relationship of biodiversity with human health, focusing on the association with mental health. A special part of the chapter will be the explanation of the underlying theories that give support to the plausible association between biodiversity and mental health. The particular mental health problem being analysed and explained is depression. Along the chapter, all relevant information for the association of biodiversity with depression will be referred as to what is found or being done in Mexico. The chapter will end with a conclusion about the need for the conservation of the different forms of biodiversity, not only for aesthetic purposes but for the positive impact on human health, despite the gaps in attributing causal effects.

2. Biodiversity

The association between physical environment and health has been known for a long time. In fact, the health and disease process is the result of a permanent interaction of human beings with the environment where they live [10]. The living and physical components of the environment, and the relationships that take place among them, define a particular ecosystem which, when it is disturbed, may produce direct and indirect alterations to the entire set of integrating elements [11]. An ecosystem then, is a complex dynamic group of various living organisms acting as a whole functioning unity [12]. The diverse group of ecosystems, the species living within those ecosystems and the genetic variations within each population, in addition to the process involving their functioning, constitutes what is called biodiversity [13].

Biological diversity or biodiversity refers to the sum of the total biotic variability present in any ecosystem; therefore, it may be estimated in different ways. Although the most common measure is by counting the number of species identified within a time and space frame (known as species richness), there are also other forms of biodiversity measurements. The multidimensional aspect of the concept allows the quantification of biodiversity using three non-exclusive criteria: a) species richness (numeric values of abundance), b) the evenness of their spatial distribution (using biodiversity index), or c) the phenotypic differentiation and genetic variability of the living organisms (at different taxonomic levels) [14]. Approximately, 1.75 millions of species have been identified in the planet, but it is estimated that the real number could be 10 times higher [15]. Ecosystems provide the supporting vital systems for any form of life on Earth, including humans. Not only provide resources for nourishing and fuel, but also they permit the air and water purification, clear and retain toxic substances, degrade waste and recycle nutrients, allow natural and crop pollination, improve soil fertility, buffer out climate change effects, among many other functions and services [1].

With more than 81,000 identified species, and a vast heterogeneity of terrestrial and aquatic ecosystems, Mexico is placed fourth world-wide in biodiversity records. Closed to 10% of the Planet biodiversity lives in Mexican territory, ranking first in reptile diversity and second in mammals, sharing with Brazil the first place in number of ecosystems [16]. In an attempt to estimate the number of species of different taxa (e.g. plants, angiosperms, amphibians, reptiles, birds, mammals, etc.) R Mittermeier created a list of the 17 countries in the world with the greatest diversity, which represents less than 10% of the Planet's surface but host seven out of ten recognised species (Table 1).

CONTINENT	COUNTRIES
Africa	Congo, Madagascar, South Africa
Asia	China, India, Indonesia, Malaysia, Philippines
Australia	Australia, Papua New Guinea
America	USA, Mexico, Brazil, Colombia, Peru, Venezuela, Ecuador

Table 1. Megadiversity countries and the Continents where they are located[17].

Biodiversity, as an important feature of ecosystems, may threaten the continuity of any form of life within when it is affected or diminished. It is estimated that 27,000 species of living organisms are lost annually (about one specie every 20 minutes), which is high above the expected rate of 3 species per year [18]. The global environmental impact due to biodiversity loss has been extensively addressed, but only recently, the focus has been centred on the health consequences of biodiversity loss.

2.1. Biodiversity and health

Human health relies in many ways on biodiversity conservation [19]. When biodiversity is affected, the entire ecosystem destabilizes reducing its resilience capacity, altering the abundance and distribution of living organisms and modifying the interactive relationships among them and with the physical environment as well. In addition, the productivity of the ecosystem is also affected, reducing the benefits that products and services may provide to humans, such as drug biosynthesis from plants and animals [20]. When natural areas are deforested for agricultural use purpose or for new urban settlements, human population becomes exposed to many vectors and species carrying communicable diseases, while limiting the population of natural predators that could exert control over the dispersion of pathogen populations [21].

The main relationship between biodiversity and human health is food provision. However, biodiversity also has direct influence on human health through other pathways not linked to food production [22]. This type of benefits has been observed in urban green spaces, where people reported more psychological benefits and better recovery capacity of mental fatigue as they were exposed to green areas with greater plant diversity [23]. The study conducted in Sheffield, UK, estimated biodiversity as species richness measured by the Gotelli-Colwell index of species density for plants. Total plant richness was the logarithmic-transformed sum estimates for woody and herbaceous plants. Butterflies and bird species were also monitored within the green space, covering a surface of 13 km². Psychological well-being was measured by the administration of a questionnaire to 312 peasants about green space usage for cognitive restoration, positive emotional bonds and sense of identity. The study found that exist a direct positive association between psychological well-being and the extension of the green space, but the association was even stronger as biodiversity increased in the green space, independently from their area sizes.

The potential benefits of biodiversity to physical and mental health have been associated mainly with direct contact of people exposed to natural environments and to the presence of urban green spaces [24]. Figure 1 exhibits the places where studies have been reported world-wide about the association between mental health and green spaces. On the other hand, the urbanisation sprawl experienced by most of the countries world-wide prevent people from open and permanent contact to natural environments. This isolation could be related to an increased number of diseases associated with urban pollution, sedentary lifestyle, and the automobile traffic overflow [25]. Therefore, all the economic and technological advantages of living in urban settings, become trades-off that jeopardise human health by modifying the environmental conditions where people live and socialize. In reference [26], it is postulated that real progress in public health will only be possible from a more humane and ecological perspective. This approach should be rooted as two fundamental dimensions of public health, that is, capable of reducing social and health inequalities and at the same time promoting health-sustaining environments. In a classical clinical study [27], it was found that surgical patients recovered faster and required less use of pain-relief medication when they could see trees outside from their room windows, as compared to a control group that only could see the walls of neighbour buildings through the windows in the hospital rooms.

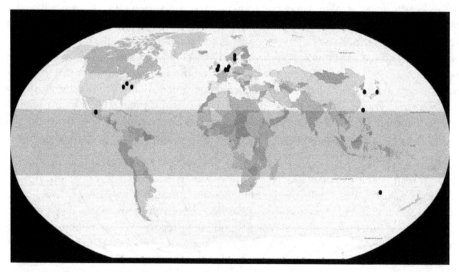

Figure 1. World mapping of the distribution of places were studies about the association of natural environments and green spaces with mental health have been conducted (according to references [2,30,82]).

In another study conducted in the Netherlands [28], it was found that people living in greener areas reported having less illness symptoms, and in general had better self-perception of their health status, including mental health. In such study, the separated effect of urban green spaces, agriculture space and natural environments were analysed, finding the strongest associations of the overall health status improvement with agricultural space living. According to the authors, this feature may reflect a Dutch condition not necessarily shared by other countries, where the green surface in agricultural areas is proportionally greater than in the other two settings. The results of this exploratory study suggested that the adults exposed to more green space (e.g. housekeepers and the elderly) report fewer symptoms, especially as the educational level increases. Recently, the same research team analysed the Dutch National Survey in General Practice to verify if the positive association found between green space exposure and good health status persisted after medical diagnoses [29]. The results indicated that not only the prevalence of 15 different group of diseases medically diagnosed were lower in residential areas with more green space (measured as the percentage of green space distributed around 1 km from the individuals' place of residence), but such an association was stronger for depression and anxiety.

For some authors, the studies linking the association of green spaces and biodiversity exposure with physical and mental health are still inconclusive, especially in urban settings [30]. However, other voices are claiming more conservation efforts, whether to enhance public health or improve aesthetics, despite any conflicting evidence [31].

3. Environmental and health components

According to reference [32], there are 5 characteristics of an area or place that influence individuals' health:

a.   Physical features of the environment shared by all residents in a locality;
b.   Availability of healthy environments at home, work, school and play;
c.   Services provided (public or private) to support people in their daily lives;
d.   Socio-cultural features of a neighbourhood; and
e.   Reputation of an area.

The first three have to do with the physical infrastructure of the place, whereas the last two are more related with the collective functioning. These categories are not mutually-exclusive and could interact, which in turn will produce different health effects on people according to their particular biological, psychosocial and economic condition. From this perspective, the study of the health effects related to living in a particular place, need to switch the traditional epidemiological paradigm that blames the individual's behaviour as the cause (or causes) of many communicable and environmental diseases. The complexity of the contexts where the health-disease processes take place, turn the conduction of etiologic studies into searching efforts at multiple time-space levels, in order to avoid the constrictions imposed by the traditional epidemiology of proximal risk factors [33]. For example, it is a myth to think that population health is better in rural environments than in urban settings only because we assume rural people is less exposed to risk factors. Studies have demonstrated that despite the health benefits of contact with natural environments, the unfavourable socioeconomic conditions of many rural people could be as an adverse as to practically wipe out any potential benefit of natural exposure [34]. It is therefore important the inclusion of the context approach where specific risk factors take place in studying population health. Those factors that modify certain health condition in a population, act differently according to the level of organization and analysis [35].

The construction and functioning of the physical and social environment of a particular area may help ameliorate or affect the health of its residents both directly and indirectly. The presence of air pollution is an example of direct effect, when airborne pollutants affect the respiratory health of individuals; whereas food provision in good quality and quantity is an example of indirect effect, when malnutrition make individuals more susceptible to any form of infectious diseases. However, individuals not always can decide the best place for their health (with better environmental quality for instance), and often the selection is indirectly determined by social and economic pre-conditions of the individual related to his or her cultural and historic background [36].

The study of the role the physical environment plays in influencing human health is a key issue in public health. According to WHO reports, environmental factors are responsible for about 24% of the total global burden of diseases [37]. In Mexico, the National Health Program 2007-2012 indicates that about 35% of the total burden of diseases is attributable to environmental factors [38]. The increased rate of species extinction along with the degradation of more than 50% of the ecosystem services world-wide jeopardises life quality and the survival of humankind [1].

3.1. Mental health and biodiversity

Although the study of the effects of contact of nature on mental health is recent, the empiric evidence exists some time ago. Authors like Erik Erikson, Harold Searle, and Paul Shepard have explained about the destruction and exploitation of nature by the so-called Western Civilization along the settlement and development of new societies, which in turn made humans more vulnerable and dependent of the emerging conditions [39]. On the other hand, there are studies focusing on the mental health effects of contact with nature in vulnerable populations [40]. In a study conducted on 112 young adults [7], it was found that the exposed group to a natural environment while doing a hike reported less anger and better humour than the group that did the hike in just urban environment. In another study, patients that were exposed to fruit smell and natural scents, reported lower prevalence of depressive episodes [41]. Animal contact has also be an alternative support method for treating psychological disorders. In reference [42], found that patients with moderate depression interacting with dolphins reported lower depression prevalence after two weeks of treatment as compared to the control group.

There are three fundamental theories (developed in the 80's last century), which try to explain the positive effect on mental health of being in contact with nature:

1. **Biophilia.-** Represents an evolution-based theory defined as the innate emotional affinity of human beings to other living organisms and nature. This feature is rooted in the hereditary aspect of human essence [43]. It is hypothesized that this behaviour is determined by a programed genetic sequence along the course of human evolution which enables a positive response to natural environments in accordance with its own survival. This theory holds that even today human beings are attracted by these natural

environments as they are perceived with a sense of "belonging" (identity) and feel they act in a more efficient manner. Reference [44] considers this nature affinity to be bound deep inside human conscience, which emerges in a similar form as other psychic experiences such as myths, poetry and religion, with a vast and complex semiotic as well. This represents the fundamentals of the moral attitude of respect to any form of life and the value of biodiversity. Based on this perspective, a new concept was developed about the affinity towards diversity (ATD), defined as the individual predisposition to appreciate the variant dynamic interaction of human and nature in the everyday situations [45]. ATD has empirically explained that future-oriented individuals and with more socializing behaviours like altruism and cooperation, tend to high rate pro-social orientation that translate into pro-environmental behaviours. Interestingly, this attitude goes beyond passive acceptance or tolerance, but includes an emotional component that expresses the preference for nature, a sense of guilt for natural resource deterioration and discomfort for actions taken by individuals or companies affecting the environment [46].

2.   **Attention-Restoration.-** This theory is based on the works of US psychologist W James at the end of the 19th century. According to this theory, in all individuals there are two areas of mental attention, a) direct attention, which is voluntary an intentional, i.e., one concentrates on aspects regarded as important for oneself. Other less important issues are classified as distractions and have to be blocked by the mind, which in turn produces mental fatigue (direct attention fatigue, or DAF); b) indirect attention (called fascination) which is involuntary and automatic, keeping concentration with low or no effort at all. This allows the brain to recover (or restore), before going back to direct attention [47]. Attention-restoration process takes place in the right side of the frontal cortex of the brain, which from an evolutionary standpoint, being alert and focused was necessary for survival. Natural environments provide the best conditions for restoration, as it allows staying away from daily routine, provide opportunity for fascination and pleasure, a sense of openness that invites the individual to explore, and the compatibility of the natural offering to one's own expectations. Moreover, just by observing a natural landscape may help restore the brain before moving to any direct attention [48].

3.   **Psycho-physiologic stress recovery.-** This theory is based on the empiric results observed in the positive responses given by individuals exposed to natural environments [49]. According to this theory, the evolution-based ability of humans to recover from a dangerous situation was a natural selection factor that increased the probability for survival. Under stressful conditions, an individual react following a physiologic mechanism pattern known as the "fight or flight response" [50]. This reaction involves catecholamine secretion (including epinephrine) into the bloodstream, which causes muscular tension, rise blood pressure, accelerates pulse rate, constrict blood vessels and increase perspiration. Thus, an individual is prepared to respond adequately when facing a fatal situation, but can restore back to its original levels once the danger has disappeared or being controlled. Some studies have found that contact with nature causes people to lower their stress level, even at a short time after the

exposure has begun. The theory considers such a response due to a limbic-associated inherently reaction of the brain (a part even more ancient than the cortex), which enabled fit individuals to have greater chances of survival during the course of human evolution [51]. In a similar way as biophilia, genetic plays a crucial role in the development of this theory.

These three theories are still under development incorporating new findings of upcoming studies. Restorative theories (attention-restoration and the stress recovery) try to explain the mechanisms by which the brain may recover after a stressful episode or mental fatigue. The main difference between the two is that the former is a more voluntary mechanism that affects the cognitive process (brain cortex), and it is measured by psychological methods, whereas the latter is more an involuntary reaction involving primitive parts of the brain (limbic system), which is measured physiologically [52].

In summary, when there is a "disconnection" of the natural world where humans live and co-exist, many diverse psychological symptoms arise including anxiety, frustration and depression, which cannot be attributed only to intra-psychological or family driven issues. It has been observed that the contact with such natural world, by means of gardening practices, animal petting, green walk or green exercise, not only relief people from depressive symptoms, but increases human capacity to be healthier, strengthen self-esteem, promotes socializing and makes people happier [53]. Although the positive association between natural contact and mental health has been consistently reported, still remains a challenge determining "how close" this "green contact" should be most appropriate [29].

4. Depression

Depression is a frequent mental disorder that currently affects life quality not only of adults, but of younger people like teenagers and children world-wide [54]. It is characterized by an overall depressed mood, with a loss of interest and/or the inability to feel anymore pleasure for things or situations that formerly produced it, loss of self-confidence and a sense of uselessness [55].

Depression diagnostics is based mostly on self-reported symptoms of patients and on clinical observations, taking as standard criteria the Diagnostic and Statistical Manual of Mental Disorders (DSM-IV-TR[1]) of the American Psychiatric Association (APA). This diagnostic tool was designed to be used with populations in different clinical settings, and represents a necessary tool to collect and communicate statistical information for public health with higher precision [56]. DSM-IV-TR provides a rather descriptive nosology than etiological approach, because it relies more on severity patterns and symptoms duration than in the inferences about the causes of the patient's disorder. DSM-IV-TR uses a multi-axial classification for a complete and systematic assessment of the different mental disorders and medical illnesses, psychosocial and environmental problems and the level of activity. Depression belongs to Axis I clinical disorders as mood disorder, which in turn are classified in depressive disorder (unipolar depression), bipolar depression and two other

---

[1] New version (DSM-V) is expected to be ready by mid-2013 according to APA.

disorder based on etiologic causes (mood disorder caused by other diseases and mood disorder caused by drug use). All depressive disorders (e.g. major depressive disorder, dysthymia, and unspecified depressive disorder) can be distinguished from bipolar disorders because there is no history of previous maniac, mixed or hypo-maniac episodes. In general, unipolar depressive disorders are more prevalent than bipolar cases [57].

4.1. Risk factors

In most patients, depressive episodes occur due to a combination of genetic, biochemical and psychosocial factors. According to [58], factors associated with depression and anxiety in the elderly may be classified in those of biological, psychological and social origin. Among those of biological origin are concurrent chronic diseases, especially cardiovascular (high and low blood pressure), cerebrovascular and psychiatric; atherosclerosis, sleeping disorders, low activity level, obesity, hearing or vision impairment, alcohol consumption, tobacco and drug use, and in general with a poor health condition. Among the risk factors of psychological origin are personal traits such as neuroticism and the history of psychiatric disorders. Finally, social risk factors identified for old individuals are low level of socialization, small and scarce social networking, living alone (no partner or spouse), problems with partner or spouse, partner or spouse on depression, low social support, parental overprotection during infancy, stressful life events in infancy, constant victim of violence, aging, among others. In [59] were identified certain consistent risk factors that suggest at least in part, they are probably causally related to the development of a major depressive disorder, and are being female, having had stressful life events, adverse experiences during childhood (e.g. physical violence, parental absence, dysfunctional family, etc.) and certain personality traits. However, the list does not include genetic vulnerability that predisposes individuals to major depressive episodes, nor the severity of such symptoms in the wide variety of depression forms.

4.2. Epidemiology

Point prevalence of depression world-wide is 1.9% in men and 3.2% in women, while for a 1-year period is 5.8% and 9.5% respectively [60]. In USA, life-time major depression prevalence is estimated to be 10.4% in non-Hispanic whites and 8.0% in Mexican-Americans, but when depression is rather moderate and chronic (i.e. dysthymia), the order reversed probably due to the low socio-economic and education levels [61]. According to the Mexican National Assessment Performance Survey (ENED), major depression in Mexico has a global prevalence of 4.5%, having women more than double of men's prevalence (5.8% vs. 2.5%). It was observed that depression prevalence increased with age but decreased as school level of individuals raised [62]. It is noteworthy that ENED reported that major depression prevalence among women is the same in rural and urban settings, whereas in men, prevalence was higher in rural environments than in urban locations. In addition, no defined pattern could be observed in the distribution of major depression among the 32 Mexican states for men and women.

4.3. Seasonal Affective Disorder (SAD)

There is growing evidence that certain mental health problems develop only during autumn and winter season, remitting on warmer and sunny seasons [63]. In USA, between 4% and 6% of adults experience SAD, while 10% to 20% develop mild forms of the disease at the end of the fall season and beginning of winter [64]. Possible causes have been linked to ocular problems to process daylight and to a deficient melatonin secretion in patients that alters their sleep-wake circadian rhythms [65]. Other studies have shown that SAD is probably also associated with problems in serotonergic transmission, since patients under white light exposure treatment responded favourably [66]. Therefore, SAD could be a morbid condition affecting countries with longer winter seasons, even though the association not necessarily is entirely latitude-dependent, and other risk factors such as genetic susceptibility and socio-cultural context could also be playing important roles [67]. Recently, it has been argued the need to consider SAD as a well-defined psychology disorder, since DSM-IV-TR is still classifying it as a cyclic effect modifier in patients with mood disorders [68].

5. Problem statement

World Health Organisation establishes that it is not possible to improve health without including mental health, because it is a fundamental aspect for life quality [69]. If no action is taken, depression is estimated to be second in disability adjusted life years (DALYs) by 2020 world-wide, and will rank first in developed countries [70]. Recent calls for prevention action have set depression as a global priority, considering not only the burden of the disease in terms of treatment cost, but on the loss of productivity as well [71]. The implementation of preventive measures to treat any disease is always desirable over the usually costlier and bothersome curative methods [52]. However, it remains unknown what is the most effective strategy to reduce depression prevalence; it is still necessary to bear in mind that prevention is one of the first goals of public health. Although there is an increasing research production aiming at studying the association between biodiversity and mental health, it is unknown the existence of specific studies in low-to-medium income countries that focus on contextual determinants associated with depression.

Depression is one of the most important diseases among Mexican adults, being the second mental disorder reported in urban settings, just after alcohol consumption [72]. Some conditions of vulnerability were identified associated with major depressive episodes, such as aging, being women, having low educational levels, and living in socioeconomically deprived areas. There is no doubt about the association between the stressful urban way-of-life and depression in adulthood. However, in the Mexican National Assessment Performance Survey (ENED-2003), data showed the same prevalence of depression symptoms between urban and rural women, but was even higher in rural men than in city men dwellers. In addition, depressive symptoms prevalence distribution per political division (State) was different between men and women, with no clear geographical pattern [62]. For men, the States with the highest prevalence were Jalisco (5%), Veracruz (4.6%) and Tabasco (4.5%), whereas the last two in the list were Nuevo León (less than 1%) and Nayarit (less than 1%). In women, Jalisco was also high with (8.2%) just after Hidalgo (9.9%) and before Estado de México (8.1%). The Mexican States with the lowest prevalence of depressive symptoms in women were Campeche (2.9%) and Sonora (2.8%). In Table 2 is possible to see the results of the total prevalence of affective disorders (including depression) for each of 6 geographic zones identified in Mexico [73]. Of note are the lowest prevalence rates registered in the South-eastern states, where biodiversity and economic deprivation are high [74, 75]

Affective Disorder	Northwest	North	Central West	Central East	Southeast	Metropolitan Areas	TOTAL
Anytime	8.4 [1.6]	9.0 (1.1)	10.2 (1.5)	10.6 (1.6)	5.7 (1.5)	10.4 (0.9)	9.1 (0.6)
Last year	4.5 (0.9)	4.6 (0.7)	5.6 (1.0)	4.9 (0.6)	2.2 (0.6)	5.3 (0.9)	4.5 (0.3)
Last month	2.4 (0.4)	1.9 (0.3)	2.5 (0.7)	2.1 (0.6)	0.9 (0.3)	2.2 (0.5)	1.9 (0.2)

Table 2. Prevalence of affective disorders by geographic zone in Mexico according to reference [73]. Standard error values are between brackets.

It is important to remark that not only humans are under stressful conditions. Planet Earth as a whole is jeopardised on its basic functions due to alterations in its structure, composition and resilience. The UN Convention on Biological Diversity (UNCBD) estimates biodiversity loss currently is close to one thousand times the natural extinction rate, and it is possible to rise in the upcoming years: around 34,000 plant species and 5,200 animal species are in danger of extinction [76]. The Millennium Development Goals entails conservation efforts for biodiversity under its seventh proposal "Environmental Sustainability" [77]. In addition, the conservation and promotion of health-sustaining environments is one of the new challenges of public health intervention [26]. In cases like Mexico, a mega-diversity country, the efforts are more than justified since the benefits to improve population's mental health have been demonstrated.

6. Research evidence

In a recent systematic review, the results of 25 studies analysing the association between green spaces and overall health and well-being were compared, finding positive consistency between exposure and some mental health-related emotions [2]. In Table 3 can be observed the results of the study where the effects of the before-and-after exposure to natural environments were compared among individuals. Consistency of results was lower when the variables were physiologically measured. These meta-analytical findings provide high internal validity to the plausible association; however, the lack of context variability (whether physical or social environments), could limit the external validity of the results. This is what reference [78] calls *psychologistic fallacy*, where individual-level studies lack the inclusion of contextual variables that may explain the apparent variability observed.

OUTCOME	EFFECT SIZE	95% CI	No. STUDIES	RESULTS
Attention	0.23	(-0.30, 0.76)	3	No effect
Energy	0.76	(0.33, 1.22)	5	Improved
Anxiety	0.52	(0.25, 0.79)	6	Improved
Tranquillity	0.07	(-0.42, 0.55)	7	No effect
Anger	0.35	(0.07, 0.64)	6	Improved
Fatigue	0.76	(0.41, 1.11)	4	Improved
Sadness	0.66	(0.66, 1.16)	3	Improved
Diastolic BP	0.32	(-0.18, 0.82)	3	No effect
Cortisol	0.57	(-0.43, 1.57)	4	No effect

Table 3. Results of the effect size (*Hedges g*) of the studies that measured health status before and after exposure to natural environments. OUTCOME = Psychologic/Physiologic variable measured. EFFECT SIZE = Group measure (Hedges g). 95% CI = 95% Confidence Interval. RESULTS = Interpretation of statistical results

Most research on the aetiology of depression and its treatment, have focused on identifying individual risk factors [79]. From a public health perspective though, it is still desirable to keep efforts on preventing the occurrence of depression rather than only in improving diagnostics and treatment efficiency [52]. In a review including more than 30 randomised control trials, it was demonstrated that different preventive interventions can reduce the incidence of major depressive episodes by as much as 50% [80]

In another systematic review of 28 studies [81], the association between physical and social characteristics of the neighbourhoods and depression in adults was analysed. The study found evidence of the negative effects of economic deprivation and the protective effect as this economic condition improved. On the other hand, the association between physical environment and depression was less evident, probably due to the few studies that incorporated the physical dimension of the neighbourhoods. Therefore, socioeconomic characteristics of higher levels of aggregation (such as individuals' place of residence), have a demonstrated effect in the mental health and well-being of the exposed population, acting independently or as effect modifiers of individual risk factor (Figure 2), but this association is less clear with the physical attributes of the environment.

In an ecological study of the association of depressive symptoms prevalence and some biodiversity indicators (measured as non-aquatic animal and plant species richness and green areas) in Mexico, it was observed that at an aggregate-level of analysis, biodiversity was positive related to depressive symptoms [82]. In other words, the study suggests that as biodiversity increases (measures as all non-aquatic species richness) in a state, depressive symptoms increase as well. For this study, data analysed were obtained from different sources. The outcome set of depressive symptoms was taken from the Mexican National Health and Nutrition Survey, ENSANUT-2006 [83]. ENSANUT-2006 was a cross-sectional survey with a probabilistic, multistage, stratified and clustered sampling. The survey collected data from October 2005 through May 2006 on health and nutritional status of the Mexican population, health services quality, public health policy and programmes, and health expenditures of Mexican dwellers [84]. The survey's structure allows representative estimations to the national, state and local levels, for urban and rural areas defined according to the population size (rural settings with less than 2,500 inhabitants; urban settings from 2,500 up to 99,999 inhabitants; metropolitan areas from 100,000). Depressive symptoms in adults were defined as those of men and women aged 20 to 65 years old, who declared having at least 5 of the following symptoms during most of the day for a period of at least one-week (DSM-IV definition of major depressive episode establishes such symptoms over a period of two weeks, therefore, we kept the focus rather on depressive symptoms only): 1) depressed mood; 2) markedly diminished interest or pleasure in almost all activities; 3) significant changes in appetite or weight; 4) insomnia (or hypersomnia in some cases); 5) psychomotor agitation or retardation; 6) fatigue; 7) feelings of worthlessness; and 8) diminished ability to think, concentrate or make decisions [56].

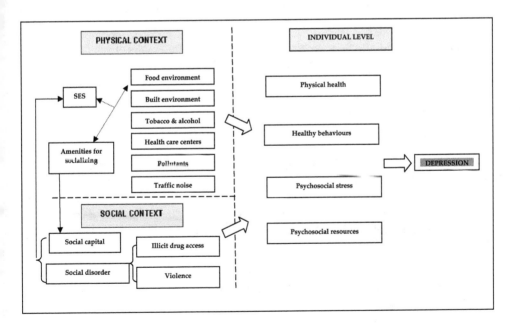

Figure 2. Neighbourhood contextual and individual risk factor model for depression in adult (modified from [81]).

The conditions were built from ENSANUT 2006 aggregating up to a state level the proportion of women, average age, and proportion of self-described as native-indigenous people. From the Mexican Compendium of Environmental Statistics 2008 [85], information was extracted per state for several biodiversity indicators such as animal (non-aquatic) and plant species richness, proportion of reforested land, proportion of natural protected areas, agricultural area and livestock grasslands. We based our ecological measure of biodiversity on the guidelines suggested by the Organisation for Economic Co-operation and Development for building environmental indicators [86].

Natural protected federal area proportions were determined according to the number of states included within its limits. The territory surface area for each state was used to calculate the proportion of green spaces occupied by re-forested, agriculture area-livestock grassland, and natural protected areas in all Mexican states. Animal and plant species richness were summed to account for the total species richness (non-aquatic biodiversity). In addition, economic disparity values were taken as Gini coefficients from the National Population Council [75], as well as the deprivation index data base. Both are measures of unfavourable socioeconomic conditions at group level, the former as an index of income-distribution inequality (the higher the value, the higher the inequality), whereas the latter measures the level of poverty based mainly on education and living conditions. Drug, tobacco and alcohol use data were obtained per state from the Mexican National Addictions Survey 2008 [87], whereas aggregate insecurity perception of individuals in every state was taken from the Mexican National Survey on Insecurity [88].

These unexpected findings are somehow in agreement with the results of a similar study in which a negative association of biodiversity with life expectancy at birth (LEB) was observed in Mexico [89]. Such eco-epidemiological study used 50 environmental indicators with information about demography, housing, poverty, water, soils, biodiversity, forestry resources, and residues were included in an exploratory factor analysis. Four factors were extracted: Population vulnerability/susceptibility, and biodiversity (FC1), urbanization, industrialization, and environmental sustainability (FC2), ecological resilience (FC3), and free-plague environments (FC4). Using ordinary least-squared regressions, it could be observed that whereas FC2, FC3, and FC4 were positively associated with life expectancy at birth, FC1 (biodiversity component) was negatively associated (Table 4). The results showed a South to North gradient inverse to the tendency with LEB. The author recommended including the physical environment as important macro-determinant when studying Mexican population health.

In another study conducted in USA, all-cause mortality in 47 largest USA cities was found to be higher in those having more green spaces [90]. They conclude that it is important the kind of contact that urban residents may have with their natural environment and the form of the green spaces as well, in order to expect the health benefits to the population, otherwise, the sprawling characteristics of USA cities may distort the positive association.

Variable	Total population			Men			Women		
	β	IC95%		β	IC95%		β	IC95%	
FC1	-0.71	-0.76	-0.64	-0.80	-0.88	-0.72	-0.62	-0.68	-0.56
FC2	0.14	0.07	0.21	0.14	0.06	0.22	0.13	0.07	0.19
FC3	0.07	0.00	0.14	0.09	0.01	0.17	0.06	-0.00	0.12
FC4	0.09	0.02	0.16	0.11	0.03	0.19	0.09	0.03	0.15
Adjusted r^2	0.9376			0.9344			0.9393		

FC1: population vulnerability/susceptibility and biodiversity.
FC2: urbanization, industrialization and environmental sustainability
FC3: ecologic resilience.
FC4: environments free of forest plagues.

Table 4. Impact of environmental factors on life expectancy at birth estimated with multiple linear regression models.

7. Conclusions

CE Winslow stated in 1920 that one of the goals of public health was prevention. Nowadays, the number of goals has increased as the population health becomes an emergent property of complex systems [91], but certainly prevention is still at the first place in the list. The challenge would be to find evidence-based effective preventive interventions [92]. Currently the relationship between biodiversity or green spaces and human health is not clear. The bulk of available evidence relating natural environments (with more biodiversity than built environments) and positive health outcomes is mainly based on data from regions with higher income and more development, which are not representative of heterogeneity of countries with less economic and human development. Studies from Latin American countries, Asia and Africa are urgently required to have a full understanding of the relationship, because there is evidence obtained in studies on other determinants of health suggesting a selection bias when data of countries with different levels of economic and human development are not included [93]. The limited evidence from developing countries, as Mexico, on biodiversity and depressive symptoms [82] and life expectancy at birth [89] is contrary to the findings in developed countries. Possible explanations to this difference include the high correlation between social determinants as income inequality, social capital, and level of democracy.

Despite of this methodological limitation to understand the causal relationships between biodiversity and depression, another plausible explanation can be related with latitudinal differences, because biodiversity decreases in regions distant from the tropics, thus, exposure to natural environments can exaggerate the positive effects. Some studies report an association between latitude and affective disorders [94-97]. An alternate explanation is related with the unit of analysis, because results of individual-level studies not necessarily are the same as those observed when analysing populations [98]. Favourable effects of biodiversity on health only have been observed in individual-level studies, whereas adverse effects have been reported in population-level studies [90]. These kinds of results are not surprising, and they are consequence of inherent limitations of science. Epidemiological and psychological studies are unable to detect the effects when low variability is present among the individuals or populations included in the studies since these approaches are based on the comparative methods.

In conclusion, we suggest that exposure to biodiversity can be good for health if the individuals are in built environments with adequate social conditions. These characteristics are frequent in Northern-European and North American countries. In contexts with higher biodiversity, the results can be ambiguous depending of the type of urbanisation [99]. As a consequence, more research in these regions is required because characteristics of the physical environment can be directly or indirectly correlated with social determinants. On the other hand, since different results are observed when studies are with individuals or populations, it is needed to include both approaches in multilevel studies. The inclusion of ecological concepts and methods will be useful to improve the quality of further studies on biodiversity and human health.

Author details

Hector Duarte Tagles
Department of Ecology and Environmental & Industrial Engineering Sonora State University (CESUES), Hermosillo, Sonora, México

Alvaro J. Idrovo
Center for Health Systems Research National Institute of Public Health, Cuernavaca, Morelos, Mexico

Public Health Department, School of Medicine, Industrial University of Santander (UIS), Bucaramanga, Santander, Colombia

8. References

[1] Millenium Ecosystem Assessment (2005). Ecosystems and Human Well-Being: Biodiversity Synthesis. World Resources Institute. Washington, USA.

[2] Bowler DE, Buyung-Ali LM, Knight TM, Pulin AS. (2010). A systematic review of evidence for the added benefits to health of exposure to natural environments. *BMC Public Health* (Open Access); 10: 456.

[3] Pretty J, Peacock J, Sellens M, Griffin M. (2005). The mental and physical health outcomes of green exercise. *Int J Environ Health Res*; 15(5): 319-37.

[4] Maas J, Verheij RA, Spreeuwenger P, Groenewegen PP. (2008). Physical activity as a possible mechanism behind the relationship between green space and health: A multilevel analysis. *BMC Public Health*; 8:206.

[5] Bauman AE. (2004). Updating the evidence that physical activity is good for health: an epidemiological review 2000-2003. *J Sci Med Sport*; 7: 6-19.

[6] Barton J & Pretty J. (2010). What is the best dose of nature and green exercise for improving mental health? A multi-study analysis. *Environ Sci Technol*; 44(10): 3947-55.

[7] Hartig T, Evans GW, Jamner LD, Davis DS, Garlin T. (2003). Tracking restoration in natural and urban field setting. *J Environ Psychol*; 32(2): 109-23.

[8] Seaman PJ, Jones R, Ellaway A. (2010). It's not just about the park, it's about integration too: why people choose to use or not use urban greenspaces. *J Behav Nutr Physical Activity*; 7:78.

[9] Bromet E, Andrade LH, Hwang I, *et al.* (2011). Cross-national epidemiology of DSM-IV major depressive episode. *BMC Medicine*; 9:90.

[10] Álvarez Alva R. (2002). Salud Pública y Medicina Preventiva. 3rd ed. Manual Moderno. México, 472 pp.

[11] Ives AR & Carpenter SR. (2007). Stability and diversity of ecosystems. *Science*; 317: 58-62.

[12] Tyler Miller G. (1994). Ecología y medio ambiente. Grupo Editorial Iberoamérica,S.A. de C.V. México, 867 pp.

[13] Magurran AE. (2010). Q&A: What is biodiversity? *BMC Biology*; 8:145.

[14] Purvis A & Hector A. (2000). Getting the measure of biodiversity. *Nature*; 405: 212-9.

[15] Chivian E & Berrnstein A. (2004). Embedded in nature: Human health and biodiversity. Editorial. *Environ Health Perspect*; 112(1): A12-3.

[16] CONABIO (2008). *Capital natural de México*. Vol. I: Conocimiento actual de la biodiversidad. Comisión Nacional para el Conocimiento y Uso de la Biodiversidad. México.

[17] Mittermeier R, Goettsch C, Robles-Gil P. (1997). Megadiversidad. Los países biológicamente más ricos del Mundo. CEMEX. México.

[18] Wilson EO. (1999). *The diversity of life*. USA: WW. Norton.

[19] Daily GC, Ehrlich PR. (1996). Global change and human susceptibility to disease. *Annu Rev Energy Environ*; 21: 125-44.

[20] Alves R, Rosa IMI (2007). Biodiversity, traditional medicine and public health: where do they meet?. *J Ethnobiology Ethnomedicine*; 3:14.

[21] Chivian E & Bernstein A. (2008). *Sustaining life: How human health depends on biodiversity*. Oxford University Press. USA

[22] Wilby A, Mitchel C, Blumenthal D, Daszak P, Friedman CS, Jutro P, Mazumder A, Prieur-Richard AH, Desprez-Loustau ML, Sharma M, Thomas MB, (2009). Biodiversity, food provision, and human health. Cap. 2. En: Sala OE, Meyerson LA, Parmesan C. (eds.) *Biodiversity change and human health. From ecosystem services to spread of disease*. Scientific Committee on Problems of the Environment (SCOPE). USA, 301 pp.

[23] Fuller RA, Irvine KN, Devine-Wright P, Warren PH, Gaston KJ. (2007). Psychological benefits of greenspace increase with biodiversity. *Biol Lett*; 3: 390-4.

[24] Dean J, van Dooren K, Weinstein P. (2011). Does biodiversity improve mental health in urban settings? *Medical Hypotheses*; 76: 877-80.

[25] Frumkin H (2002). Urban Sprawl and Public Health. *Public Health Reports*; 117: 201-17.

[26] McMichael AJ, Beaglehole R. (2000). The changing global context of public health. *Lancet*; 356: 495-9.

[27] Ulrich R (1984). View through a window may influence recovery from surgery. *Science*; 224(4647): 420-1.

[28] De Vries S, Verheij RA, Groenewegen PP, Spreeuwenberg P. (2003). Natural environments-healthy environments? An exploratory analysis of the relationship between greenspace and health. *Environment and Plannig A*; 35: 1717-31.

[29] Maas J, Verheij RA, de Vries S, Spreeuwenger P, Schellevis FG, Groenewegen PP. (2009). Morbidity is related to a green living environment. *J Epidemiol Community Health*; 63: 967-73.

[30] Lee ACK & Maheswaran R. (2011). The health benefits of urban green spaces: a review of the evidence. *J Public Health*; 33(2): 212-22.

[31] Schultz N. (2010). Country vs. city: Green spaces are better for you. *New Scientist*; 2785.

[32] Macintyre S, Ellaway A, Cummins S. (2002). Place effects on health: how can we conceptualise, operationalise and measure them?. *Soc Sci Med*; 55: 125-39.

[33] McMichael AJ. (1999). Prisoners of the proximate: loosening the constraints on epidemiology in an age of change. *Am J Epidemiol*; 149: 887-97.

[34] Schenker MB. (1996). Preventive medicine and health promotion are overdue in the agricultural workplace. *J Public Health Policy*; 17(3): 275-305.

[35] Diez Roux AV. (2008). La necesidad de un enfoque multinivel en epidemiología. *Región y Sociedad*; 20(2): 77-91.

[36] Verheij RA. (1996). Explaining urban-rural variations in health: a review of interactions between individual and environment. *Soc Sci Med*; 42: 923-35.

[37] WHO (2009). Fact sheets on environmental health. World Health Organization. Available: http://www.who.int/topics/environmental_health/en/. Accessed 2009 December 02.

[38] Secretaría de Salud (2007). *Programa Nacional de Salud 2007-2012. Por un México Sano: construyendo alianzas para una mejor salud.* México, 74 pp.

[39] Shepard P. (1982). *Nature and Madness.* Sierra Club, CA.

[40] Wells NM & Evans GW. (2003). Nearby Nature: A buffer of life stress among rural children. *Environ Behav*; 35(3): 311-30.

[41] Schiffman S. (1992). Aging and the sense of smell: Potential benefits of fragrance enhancement. In: Van Toller & G Dodd (eds). *Fragrance: The psychology and biology of perfume.* Elsevier. England, pp 51-66.

[42] Antonioli C & Reveley M. (2005). Randomised controlled trial of animal facilitated therapy with dolphins in the treatment of depression. *British Medical Journal*; 331(7527), 1231.

[43] Wilson EO. (1984). *Biophilia: The human bond with other species.* Cambridge: Harvard University Press.

[44] Kellert S & Wilson E (eds.) (1993). The biophilia hypothesis. Island Press. USA, pp. 334.

[45] Corral-Verdugo V et al. (2010). *Psychological approaches to sustainability.* Nova Science Publishers.

[46] Corral-Verdugo V, Bonnes M, Tapia-Fonllem C, Fraijo-Sing B, Frías-Armenta M, Carrus G. (2009). Correlates of pro-sustainability orientation: The affinity towards diversity. *J Environ Psychol*; 29: 34-43.

[47] Kaplan R & Kaplan S (1989). *The experience of nature: A psychological perspective.* New York: Cambridge University Press.

[48] Kaplan S (1995). The restorative effects of nature: Toward an integrative framework. *J Environ Psychol*; 15: 169-82.

[49] Ulrich R (1983). Aesthetic and effective response to natural environment. In: Altman I & Wolhwill JF (Eds.). *Behavior and the natural environment.* NY, Plenum Press, 85-125.

[50] McEwen BS, Stellar E (1993). Stress and the individual: mechanisms leading to disease. *Arch Int Med*; 153: 2093-101.

[51] Ulrich R, Simons RF, Losito E, Fiorito E, Miles MA and Zelson M. (1991). Stress recovery during exposure to natural and urban environments. *J Environ Psychol*; 11: 201-30.

[52] Bird W. (2007). *Natural thinking. Investigating the links between the natural environment, biodiversity and mental health.* The Royal Society for the Protection of Birds. UK, 116 pp.

[53] Chalquist C. (2009). A look at the ecotherapy research evidence. *Ecopsychology*; 1(2): 1-11.

[54] Benjet C, Borges G, Medina-Mora ME. (2008). DSM-IV personality disorders in Mexico: results from a general population survey. *Revista Brasileira Psiquiatria*; 30(3): 227-34.

[55] De la Garza Gutiérrez F. (2004). *Depresión, angustia y bipolaridad. Guía para pacientes y familiares.* Ed. Trillas, México, 233 pp.

[56] APA (1994). *Diagnostic and Statistical Manual of Mental Disorders (DSM-IV).* American Psychiatric Association. 4th ed.

[57] Carney RM, Freedland KE. (2000). *Depression and medical illness.* Ch. 9. In: Berkman LF and Kawachi I (eds.). *Social Epidemiology.* Oxford University Press. USA, 191-212.

[58] Vink D, Aartsen MJ, Schoevers RA. (2008). Risk factors for anxiety and depression in the elderly: a review. *J Affective Disorders*; 106: 29-44.

[59] Fava M & Kendler K (2000). Major depressive disorder. *Neuron*; 28(2): 335-41.

[60] OMS (2001). Informe sobre la salud en el mundo. Salud mental: nuevos conocimientos, nuevas esperanzas. Organización Mundial de la Salud, Suiza. p. 29-30.

[61] Riolo SA, Nguyen TA, Greden JF, King CA. (2005). Prevalence of depression by race/ethnicity: findings from the National Health and Nutrition Examination Survey III. *Am J Public Health*; 95(6): 998-1000.

[62] Belló M, Puentes-Rosas E, Medina-Mora ME, Lozano R (2005). Prevalencia y diagnóstico de depresión en población adulta en México. *Salud Publica Mexico*; 47(suppl 1): S4-S11.

[63] Michalak EE, Wilkinson C, Dowrick C, Wilkinson G. (2001). Seasonal affective disorder: prevalence, detection and current treatment in North Wales. *Br J Psychiatry*; 179: 31-4.

[64] Cunningham A. (2011). Depression common with shorter days, less sunlight. USATODAY.com. Published on 2011 Jan 30.

[65] Wehr TA, Duncan WC Jr, Sher L, Aeschbach D, Schwartz PJ, Turner EH. (2001). A circadian signal of change of season in patients with seasonal affective disorder. *Arch Gen Psychiatry*; 58:1108–14.

[66] Virk G, Reeves G, Rosenthal NE, Sher L, Postolache TI. (2009). Short exposure to light treatment improves depression scores in patients with seasonal affective disorder: A brief report. *Int J Disabil Hum Dev*; 8(3): 283-6.

[67] Mersch PPA, Middendorp HM, Bouhuys AL, Beersma DGM, van der Hoofdakker RH. (1999). Seasonal affective disorder and latitude: a review of the literature. *J Affective Disorder*; 53(1): 35-48.

[68] Rosenthal NE. (2009). Issues for DSM-V: Seasonal affective disorder and seasonality. *Am J Psychiatry*; 166: 852-3.

[69] WHO (2005). *European Declaration on Mental Health*. World Health Organization.

[70] Murray CJ and Lopez AD. (1997). Alternative projections of mortality and disability by cause 1990-2020: Global burden of disease study. *Lancet*; 349: 1498-1504.

[71] Cuijpers P, Beekman ATF, Reynolds CF. (2012). Preventing Depression. A Global Priority. *JAMA*; 307(10): 1033-4.

[72] Medina-Mora ME, Borges G, Benjet C, Lara C, Berglund P. (2007). Psychiatric disorders in Mexico: lifetime prevalence in a nationally representative sample. *Br J Psychiatry*; 190(6): 521-8.

[73] Medina-Mora ME, Borges G, Lara Muñoz C, Benjet C, Blanco Jaimes J, Fleiz Bautista C, Villatoro Velázquez J, Rojas Guiot E, Zambrano Ruíz J, Casanova Rodas L, Aguilar-Gaxiola S. (2003). Prevalencia de trastornos mentales y uso de servicios: resultados de la Encuesta Nacional de Epidemiología Psiquiátrica en México. *Salud Mental*; 26(4): 1-16.

[74] SEMARNAT (2010). Compendio de Estadísticas Ambientales. Secretaría de Medio Ambiente y Recursos Naturales. México. Available: (http://www.semarnat.gob.mx). Accessed: 2012 April 02

[75] Tuirán-Gutiérrez R. (2005). La desigualdad en la distribución del ingreso monetario en México. *Consejo Nacional de Población* (CONAPO). México, 248 pp.

[76] UNCBD (2010). United Nations Convention on Biological Diversity. Available: http://www.cbd.int/2010/biodiversity/?tab=0,1,2 Accessed: 2012 April 02

[77] UN (2005). Millenium Development Goals. United Nations. Available: http://www.un.org/millenniumgoals/ Accesed : 2010 December 05.

[78] Diez-Roux AV. (1998). Bringing Context back into Epidemiology: Variables and Fallacies in Multilevel Analysis. *American Journal of Public Health* 88 (2): 216-222.

[79] Keller MB. (2003). Past, present and future directions for defining optimal treatment outcome in depression: remission and beyond. *JAMA*; 289: 3152-60.

[80] Muñoz RF, Cuijpers P, Smith F, Barrera AZ, Leykin Y. (2010). Prevention of major depression. *Annu Rev Clin Psychol*; 6: 181-212.

[81] Kim D. (2008). Blues from the neighborhood? Neighborhood characteristics and depression. *Epidemiol Rev*; 30: 101-17.

[82] Duarte Tagles H, Idrovo AJ (2012). Biodiversity and depressive symptoms in the tropics: a fsQCA of Mexican data. (Submitted)

[83] Olaiz-Fernández G, Rivera-Dommarco J, Shamah-Levy T, et al. (2006). Encuesta Nacional de Salud y Nutrición 2006. *Instituto Nacional de Salud Pública*. México.

[84] Rodríguez-Ramírez S, Mundo-Rosas V, Jiménez-Aguilar A, et al. (2009). Methodology for the analysis of dietary data from the Mexican National Health and Nutrition Survey 2006. *Salud Publica Mex*; 51(Suppl 4):S523-9.

[85] SEMARNAT (2008). *Compendio de Estadísticas Ambientales*. México: Secretaría de Medio Ambiente y Recursos Naturales.

[86] OECD (2003). *Environmental Indicators. Development, measurement and use*. France: Organisation for Economic Co-operation and Development.

[87] Rodríguez-Ajenjo C, Villatoro-Velázquez JA, Medina-Mora ME, et al. (2009). *Encuesta Nacional de Adicciones 2008*. México: Instituto Nacional de Salud Pública.

[88] De la Barreda L, Ruiz-Harrel R, Sayeg-Seade C, et al. (2005).*Tercera Encuesta Nacional Sobre Inseguridad 2005*. México: Instituto Nacional de Estadística y Geografía.

[89] Idrovo AJ. (2011). Physical environment and life expectancy at birth in Mexico: an eco-epidemiological study. *Cad Saude Publica*;27:1175-84.

[90] Richardson EA, Mitchell R, Hartig T, et al. (2012). Green cities and health: a question of scale? *J Epidemiol Community Health*; 66:160-5.

[91] Diex-Roux AV. (2007). Integrating social and biologic factors in health research: A systems view. *Ann Epidemiol*; 17: 569-74.

[92] Sackett DL, Rosenberg WM. (1995). The need for evidence-based medicine. *J R Soc Med*, 88: 620-4.

[93] Idrovo AJ, Ruiz-Rodrígucz M, Manzano-Patiño AP (2010). Beyond the income inequality hypothesis and human health: a worldwide exploration. *Rev Saude Publica J*; 44: 695-702.

[94] Mersch PP, Middendorp HM, Bouhuys AL, Beersma DG, van den Hoofdakker RH (1999). Seasonal affective disorder and latitude: a review of the literature. *J Affect Disord*; 53: 35-48.

[95] Kegel M, Dam H, Ali F, Bjerregaard P (2009). The prevalence of seasonal affective disorder (SAD) in Greenland is related to latitude. *Nord J Psychiatry*; 63: 331-5.

[96] Levitt AJ, Boyle MH (2002). The impact of latitude on the prevalence of seasonal depression. *Can J Psychiatry*; 47: 361-7.

[97] Bloom DE, Canning D, Fink G (2008). Urbanization and the wealth of nations. *Science*;319:772-5.

[98] Rose G (1985). Sick individuals and sick populations. *Int J Epidemiol*; 14: 32-8.

[99] Rosen LN, Targum SD, Terman M, Bryant MJ, Hoffman H, Kasper SF, et al. (1990). Prevalence of seasonal affective disorder at four latitudes. *Psychiatry Res*; 31: 131-44.

Permissions

The contributors of this book come from diverse backgrounds, making this book a truly international effort. This book will bring forth new frontiers with its revolutionizing research information and detailed analysis of the nascent developments around the world.

We would like to thank Gbolagade Akeem Lameed, for lending his expertise to make the book truly unique. He has played a crucial role in the development of this book. Without his invaluable contribution this book wouldn't have been possible. He has made vital efforts to compile up to date information on the varied aspects of this subject to make this book a valuable addition to the collection of many professionals and students.

This book was conceptualized with the vision of imparting up-to-date information and advanced data in this field. To ensure the same, a matchless editorial board was set up. Every individual on the board went through rigorous rounds of assessment to prove their worth. After which they invested a large part of their time researching and compiling the most relevant data for our readers. Conferences and sessions were held from time to time between the editorial board and the contributing authors to present the data in the most comprehensible form. The editorial team has worked tirelessly to provide valuable and valid information to help people across the globe.

Every chapter published in this book has been scrutinized by our experts. Their significance has been extensively debated. The topics covered herein carry significant findings which will fuel the growth of the discipline. They may even be implemented as practical applications or may be referred to as a beginning point for another development. Chapters in this book were first published by InTech; hereby published with permission under the Creative Commons Attribution License or equivalent.

The editorial board has been involved in producing this book since its inception. They have spent rigorous hours researching and exploring the diverse topics which have resulted in the successful publishing of this book. They have passed on their knowledge of decades through this book. To expedite this challenging task, the publisher supported the team at every step. A small team of assistant editors was also appointed to further simplify the editing procedure and attain best results for the readers.

Our editorial team has been hand-picked from every corner of the world. Their multi-ethnicity adds dynamic inputs to the discussions which result in innovative outcomes. These outcomes are then further discussed with the researchers and contributors who give their valuable feedback and opinion regarding the same. The feedback is then collaborated with the researches and they are edited in a comprehensive manner to aid the understanding of the subject.

Apart from the editorial board, the designing team has also invested a significant amount of their time in understanding the subject and creating the most relevant covers. They scrutinized every image to scout for the most suitable representation of the subject and create an appropriate cover for the book.

The publishing team has been involved in this book since its early stages. They were actively engaged in every process, be it collecting the data, connecting with the contributors or procuring relevant information. The team has been an ardent support to the editorial, designing and production team. Their endless efforts to recruit the best for this project, has resulted in the accomplishment of this book. They are a veteran in the field of academics and their pool of knowledge is as vast as their experience in printing. Their expertise and guidance has proved useful at every step. Their uncompromising quality standards have made this book an exceptional effort. Their encouragement from time to time has been an inspiration for everyone.

The publisher and the editorial board hope that this book will prove to be a valuable piece of knowledge for researchers, students, practitioners and scholars across the globe.

List of Contributors

Jessica P. Karia
Enviro-GIS, Makarpura, Vadodara,Gujarat, India

Rosario Tejera
Research Group for Sustainable Management, Department of Economy and Forest Management - E.T.S.I de Ingenieros de Montes, Technical University of Madrid (U.P.M), Ciudad Universitaria, Madrid, Spain

María Victoria Núñez
Research Group for Sustainable Management, Department of Projects and Rural Planning - E.T.S.I de Ingenieros de Montes, Technical University of Madrid (U.P.M), Ciudad Universitaria, Madrid, Spain

Ana Hernando
Research Group for Sustainable Management, Department of Economy and Forest Management - E.T.S.I de Ingenieros de Montes, Technical University of Madrid (U.P.M), Ciudad Universitaria, Madrid, Spain

Javier Velázquez
Catholic University of Avila. C/ Los Canteros, Ávila, Spain

Ana Pérez-Palomino
E.T.S.I de Ingenieros de Montes, Technical University of Madrid (U.P.M), Ciudad Universitaria, Madrid, Spain

Benjamin L. Allen and Luke K-P. Leung
School of Agriculture and Food Sciences, the University of Queensland, Gatton, Queensland, Australia

Peter J.S. Fleming
Vertebrate Pest Research Unit, Department of Primary Industries, Orange, New South Wales, Australia

Matt Hayward
Centre for African Conservation Ecology, Nelson Mandela Metropolitan University, Port Elizabeth, South Africa
School of Biological, Earth and Environmental Science, University of New South Wales, Sydney, Australia

Lee R. Allen
Robert Wicks Pest Animal Research Centre, Biosecurity Queensland, Toowoomba, Queensland, Australia

Richard M. Engeman
National Wildlife Research Centre, US Department of Agriculture, Fort Collins, Colorado, USA

Guy Ballard
Vertebrate Pest Research Unit, Department of Primary Industries, Armidale, New South Wales, Australia

A. G. Lameed
Department of Wildlife and Ecotourism Management, Faculty of Agriculture and Forestry, University of Ibadan, Ibadan, Nigeria

Jenyo-Oni Adetola
Department of Aquaculture and Fisheries Management, Faculty of Agriculture and Forestry, University of Ibadan, Ibadan, Nigeria

Flavia del Valle Loto
Pilot Plant of Industrial and Microbiological Processes (PROIMI), CONICET, Tucumán, Argentina

Analía Alvarez
Natural Sciences College and Miguel Lillo Institute, National University of Tucumán, Tucumán, Argentina
Pilot Plant of Industrial and Microbiological Processes (PROIMI), CONICET, Tucumán, Argentina

Thiago Metzker
Programa de Pós-Graduação em Ecologia, Conservação e Manejo da Vida Silvestre, Brazil

Tereza C. Spósito
Departamento de Botânica, Universidade Federal de Minas Gerais (UFMG), CP 486, CEP 31270-970, Belo Horizonte, Brazil

Britaldo S. Filho
Departamento de Geociências, UFMG, Belo Horizonte, Brazil

Jorge A. Ahumada
Tropical Ecology Assessment and Monitoring Network, Science and Knowledge Division, Conservation International, Arlington, VA, USA

Queila S. Garcia
Departamento de Botânica, Universidade Federal de Minas Gerais (UFMG), CP 486, CEP 31270-970, Belo Horizonte, Brazil

Yoshitaka Oishi
Department of Forest Science, Faculty of Agriculture, Shinshu University, Japan

I. A. Orellana
Andean Patagonian Forest Research and Extension Center (CIEFAP), Patagonia San Juan Bosco National University (UNPSJB), Esquel, Chubut, Argentina

E. Raffaele
Ecotone Laboratory, Comahue National University - INIBIOMA/ CONICET, S. C. de Bariloche, Río Negro, Argentina

Hector Duarte Tagles
Department of Ecology and Environmental & Industrial Engineering Sonora State University (CESUES), Hermosillo, Sonora, México

Alvaro J. Idrovo
Center for Health Systems Research National Institute of Public Health, Cuernavaca, Morelos, Mexico
Public Health Department, School of Medicine, Industrial University of Santander (UIS), Bucaramanga, Santander, Colombia

Printed in the USA
CPSIA information can be obtained
at www.ICGtesting.com
JSHW011427221024
72173JS00004B/701